LEBANESE AMBER

LEBANESE AMBER

The Oldest Insect Ecosystem in Fossilized Resin

George O. Poinar, Jr.
Raif K. Milki

Oregon State University Press
Corvallis

Publication of this book was made possible by
a contribution from the

Safadi Foundation

The Oregon State University Press is grateful for this support

Front cover photograph of a dance fly close to the genus Brachystoma *in the subfamily Brachystomatinae (Empididae: Diptera) and all photographs inside the book by George O. Poinar, Jr.*

Back cover photograph of Dr. Milki collecting amber from 135 million year old Lower Cretaceous beds on the slopes of Mt. Lebanon by Nesrine Milki

The paper in this book meets the guidelines for permanence and durability of the Committee on Production Guidelines for Book Longevity of the Council on Library Resources and the minimum requirements of the American National Standard for Permanence of Paper for Printed Library Materials Z39.48-1984.

Library of Congress Cataloging-in-Publication Data
Poinar, George O.
  Lebanese amber : the oldest insect ecosystem in fossilized resin /
George O. Poinar, Jr., and Raif Milki.-- 1st ed.
      p. cm.
Includes bibliographical references and index.
  ISBN 0-87071-533-X (alk. paper)
  1. Amber--Lebanon. 2. Amber fossils--Lebanon. I. Milki, Raif.
II. Title.
  QE391.A5 P65 2001
  560'.95692--dc21

                                    2001003008

Oregon State University Press
101 Waldo Hall
Corvallis OR 97331-6407
541-737-3166 • fax 541-737-3170
http://osu.orst.edu/dept/press

OREGON STATE
UNIVERSITY

Dedication

We dedicate this book to Mohammed Safadi,
who made its publication possible through a contribution
from the Safadi Foundation, which promotes higher
education, technology and research.

CONTENTS

Tables

FOREWORD

Some one hundred and thirty million years ago, when dinosaurs roamed the earth, towering kauri pines in prehistoric Lebanon wept copious amounts of resin. The resin, which trapped a diverse range of life, especially insects, eventually transformed into what is known today as Lebanese amber. This is the oldest known amber to preserve insect remains and possibly also contains the earliest angiosperm leaves. Entombed biting insects may even contain the blood of dinosaurs.

Amber is known for preserving fossils in life-like condition, because they were not subjected to the compression that all too commonly affects most soft-bodied organisms that enter the prehistoric record. Inclusions in amber are three dimensional and appear ready to spring out of their golden tombs and continue their former lives.

Here, for both the professional and amateur, is a well-illustrated account of Lebanese amber from the Early Cretaceous. Included are records of the first known appearances of many insect groups, all from that significant geological interval that so altered the terrestrial world—the beginning of the flowering plants.

In this first book on amber from Lebanon, the authors include information from prior descriptions of individual fossils and add a wealth of new material documented by photographs. They also provide background information on the geology and occurrence of Lebanese amber and a comprehensive section on other types of resins and gums found in the Near East that might be confused with true amber. Emphasis is placed on co-evolutionary relationships found in Lebanese amber, some of which persist to the present day. It is a pleasure to read this work and view the color plates beautifully depicting the most ancient insects from any amber source.

Dr. Arthur Boucot, Department of Zoology,
Oregon State University

PREFACE

The present work is a review of all the organisms thus far reported from Lebanese amber. Various paleoentomologists have contributed to the study of Lebanese amber insects. Studies by Paul Whalley, once at the British Museum, have been especially useful. However, works like ours are also made possible by those who go into the field and search for amber sites. Scientists are indebted to these individuals since, without their zeal, there would not be many scientific descriptions of amber fossils or books like the present one.

While there are a few collections of Lebanese amber in public institutions and museums (most of which are inaccessible to viewing), the majority of this rare fossilized resin resides in private collections. The fossils depicted in the photographs presented here were collected by Raif Milki on numerous trips into the field dating back to 1962. These rare inclusions comprise the Milki Lebanese amber collection maintained at the American University of Beirut in Lebanon.

Figure 1. Reconstruction of the Early Cretaceous Lebanese amber forest. Plants include leaves of *Agathis levantensis* with one male and two immature female cones in the upper center, leaves of the tree fern *Weichselia* in the lower right and a *Zamites* cycadophyte in the lower left. Dinosaurs include (from left to right) the head of the large carnivore *Carcharodontosaurus*; the herbivore *Ouranosaurus* with its skin sail; the swift predator *Elaphrosaurus* in the foreground; and the large sauropod *Dicraeosaurus* in the background. A pterosaur, *Pterodactylus*, glides overhead and a small triconodont mammal cowers in the left foreground. A biting midge of the genus *Protoculicoides* is entrapped in resin on the bark of the kauri tree in the upper left. (Drawing by G. Poinar)

SCIENTIFIC ASPECTS OF LEBANESE AMBER

Introduction

Amber is one of the great natural treasures of Lebanon. The scientific importance of Lebanese amber lies in its great age. This amber dates back to the Early Cretaceous and contains the oldest known arthropods of any fossilized resin deposit. These now extinct organisms lived in a forest different from any in existence today, long before the land known as Lebanon reached the Mediterranean Sea. The resin-producing woods originated in the southern hemisphere when Lebanon was part of the great continent of Gondwanaland (Figure 2).

Lebanese amber was formed in a tropical-subtropical forest consisting predominately of kauri pines, cycads, and ferns and dominated by reptiles including dinosaurs and pterosaurs (Figure 1).

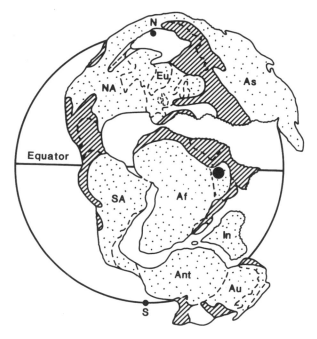

Figure 2. Arrangement of the continents in the Early Cretaceous when the Lebanese amber forest existed. Black dot represents region of the Lebanese amber forest. Af = Africa; Ant = Antarctica; As = Asia; Au = Australia; Eu = Europe; In = India; N = North Pole; NA = North America; S = South Pole; SA = South America.

This sylvan habitat contained some of the earliest flowering plants (Angiosperms), which had begun to usurp the dominating gymnosperms and spore-bearing plants. Lebanese amber may therefore hold answers to many questions about the origin of these higher plants.

All of the fossils thus far described from Lebanese amber are extinct at the species level; most are extinct at the genus level, but the majority belong to extant families. The present work includes four phyla and six classes, with insect representatives of fifteen orders, twenty-nine families, fifty-six genera, and sixty-nine species. The striking cases of four insect genera from Lebanese amber still extant today represent the longest generic lineages of any land animals.

The Early Cretaceous amber beds in Lebanon cross over the boundaries of this small country into neighboring areas. This is why these deposits are sometimes known as "Middle East," "Near East," or "Levantine" amber. However, since the vast majority of fossiliferous material so far collected and studied has come from Lebanon, the deposits are most frequently referred to as Lebanese amber.

Geological Setting

The present geographical position of Lebanon has changed from its location some 150 million years ago, when this area was part of the supercontinent Gondwanaland (Figure 2). Lebanon then was part of the Arabian Peninsula attached to the eastern portion of the African continent and situated on the equator. Africa was still connected to South America as part of Gondwanaland. India and Madagascar were united in the southern hemisphere and had only recently separated from the landmasses of Antarctica, Australia, and New Zealand.

The period of approximately 150 to 100 million years ago was one of uplift and volcanic activity, eventually followed by erosion which resulted in the formation of rivers, swamps and deltas. At that time , the region that we now know as Lebanon was a deltaic plain that alternately rose above and fell below sea level, thus accumulating large amounts of marine limestone. Remnants of these deposits occur in the gorges of northern Mount Lebanon and in the mountain ranges of Barouk and Hermon. It was during this era that the resiniferous forests responsible for Lebanese amber existed. The resin fell from the trees, hardened, and was washed into low-lying plains that were later covered by shallow seas. Slowly the amber was deposited in sands and shales of great thickness.

Figure 3. Lebanon (major cities designated by squares) with amber localities (designated by circles) on the western slopes of Mount Lebanon.

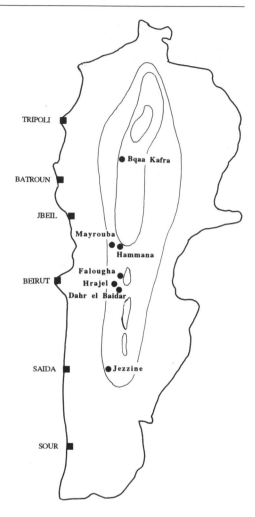

Some 100 to 50 million years ago, long after the amber was formed, limestones and chalks containing the well-known Lebanese fish fossils were deposited. Over the past twenty million years, the African continent, with its attached Arabian Peninsula (including Lebanon), collided with Eurasia, uplifting the rock layers containing the amber to form the Lebanese mountains (Smith et. al., 1994). Wearing away of the wetter western slopes of this range eventually exposed the Early Cretaceous amber-bearing beds where most of the deposits occur (Figure 3). Localities range from Jezzine in the south to Bqaa Kafra in the north. It is quite likely that amber is present on the eastern slopes of these mountains as well as in the Anti-Lebanon ranges but buried beneath thick deposits.

Age of Lebanese Amber

There have been differing opinions regarding the age of Lebanese amber, partly due to changes made in assigning dates to the various stages of the Cretaceous. The last report by Schlee (1990) gave a range of 130 to 135 million years, which corresponds to a period from the Barremian to the Hauterivian or beginning of the Valanginian (Harland et al., 1990). While some amber is found in younger Albian-Aptian beds (97-124 million years), these are considered secondary deposits (Schlee and Dietrich, 1970). The previously reported age of 130 to 135 million years is accepted here.

All indications are that amber was being formed in the neighboring countries at approximately the same time. Jordanian amber is now believed to be roughly the same age, considerably older than the Aptian-Albian age, as previously thought (Shinaq and Bandel, 1998). In addition, amber from Mount Hermon, the eastern escarpments of the Naftali Mountains, and other areas south of Lebanon throughout the Levant is considered Hauterivian-Valanginian in age (Nissenbaum and Horowitz, 1992). Thus the present evidence suggests that the forest of resin-bearing trees covered a large portion of the Arabian Peninsula in the Early Cretaceous. The amber is found in sandstone and limestone sedimentary layers (Plate 1), as well as in lignitic beds among the sedimentary layers (Plate 7), most originating from the Grès de Base Formation in Lebanon (Schlee and Dietrich, 1970)(Poinar, 1992) and the Kurnub sandstone Formation in Jordan (Shinaq and Bandel, 1998).

The Study of Lebanese Amber

During the millions of years Lebanese amber has been in the earth, the sedimentary layers containing it have been subjected to various stresses as a result of earth-moving forces. This, coupled with the regular daily and seasonal fluctuations in temperature affecting the surface layers, has modified the color and character of the amber. Thus it is not surprising that much of this aged material is highly fractured and tends to fall apart as it is removed from the earth, making collection a tedious and delicate task (Plate 1). Most of the amber is collected in small pieces less than a centimeter in diameter, though some pieces are fist sized and quite durable (Figure 4). It is difficult to clearly see many of the inclusions in the amber matrix; studying and photographing insect specimens in Lebanese amber is quite a challenge.

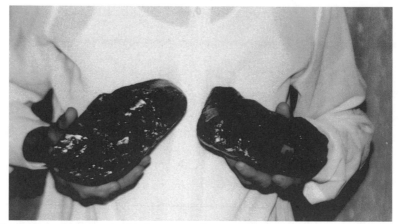

Figure 4. Largest pieces of Lebanese amber yet documented.

There are several ways of preventing further deterioration of the amber. Perhaps the best method involves embedding the amber pieces in liquid plastic. The plastic enters the cracks and helps to clarify the amber as well as protect it from further environmental effects. After the plastic has hardened, both the plastic and embedded amber can be polished without fear of fracturing the amber further. Care must be taken with this method however, since if the liquid plastic makes contact with the membranous surface of insect wings, it can obscure the fine venation and mask important taxonomic characteristics.

Lebanese amber varies in color from transparent light orange (Plate 3) to dark brown and even black (Plate 4). Much of the amber is opaque (Plate 5), sometimes with circular (Plate 6) or longitudinal layers. Some of the amber may be embedded in highly carbonized strata (Plate 7). The lighter transparent pieces are most valuable to paleontologists since the fossils can be observed with fewer obscurities.

Plant Source

Modern analytical methods have been used to identify the Lebanese amber plant source. One procedure, known as carbon-13 nuclear magnetic resonance (13 CNMR) spectroscopy (abbreviated here to NMR), characterizes ambers by matching their spectra with those of resin from living trees. This method is based on the proven assumption that, even after millions of years, chemical compounds in fossilized resin are little modified and can be matched with compounds in resin from living trees. Using this type of "fingerprinting," Lebanese amber, as well as similarly aged amber from neighboring countries, was determined to have been produced by coniferous trees of the family Araucariaceae, in particular members of the genus *Agathis* (Lambert et al., 1996). This tree genus, one of three extant genera in the Araucariaceae and commonly known as kauri pines, is absent from the northern hemisphere today. However chemical analyses of Cretaceous and even early Tertiary ambers from different localities, show how widespread kauri pines were throughout the northern hemisphere (Table 1). Why these trees exist only in the Australasian region of the southern hemisphere today (Table 2) is an intriguing question.

An *Agathis* source for Lower Cretaceous amber from Jordan and Lebanon was also reported by Bandel and Vavra (1981), who analyzed fossilized resin with infrared spectroscopy, mass spectroscopy, and thin layer chromatography. While these authors mentioned finding some differences between Lebanese and Jordanian amber, the NMR spectra of amber from these two sources were identical (Lambert et al., 1996).

Coalified *Agathis* leaf impressions commonly occur in amber-bearing sandstones and siltstones of the Kurnub group in the Zerka Valley near the village of Khirbat es Suweirat in Jordan (Bandel and Haddadin, 1979; Bandel and Vavra, 1981). It is highly likely that these leaves originated from the trees that produced the Near Eastern amber. A description of the amber *Agathis* tree based on these leaves and other characteristics is presented in the Appendix. This new species, *Agathis levantensis* sp. n. is considered the source of the Early Cretaceous amber of Lebanon and the surrounding territories.

Figure 5. Pair of fossilized leaves of *Agathis levantensis* from amber bearing Early Cretaceous sedimentary deposits in the Zerka valley of Jordan.

Table 1. Evidence of extinct kauri pines (*Agathis* spp.) in the northern hemisphere based on resin analysis

Location	Date of deposit (mya)	Types of analysis	Reference
Baltic region (Europe)	40	Infra red spectrometry, Chemical analysis	Langenheim, 1969; Thomas, 1969; Gough and Mills, 1972
Washington State (USA)	50	Nuclear magnetic resonance	Lambert et al., 1990, 1996
British Columbia (Canada)	55	Nuclear magentic resonance	Poinar et al., 1999
Alberta, Manitoba (Canada)	70-80	Infra red spectrometry	Langenheim and Beck, 1968
		Pyrolysis mass spectrometry	Poinar and Haverkamp, 1985
		Nuclear magnetic resonance	Lambert et al., 1990, 1996
Kansas (USA)	70-80	Nuclear magnetic resonance	Lambert et al., 1996
Mississippi (USA)	80-90	Nuclear magnetic resonance	Lambert et al., 1996
New Jersey (USA)	70-95	Nuclear magnetic resonance	Lambert et al., 1990
France	90-97	Nuclear magnetic resonance	Lambert et al., 1996
Alaska (USA)	100	Pyrolysis mass spectrometry	Poinar and Haverkamp, 1985
		Nuclear magnetic resonance	Lambert et al., 1990, 1996
Greenland	104-112	Nuclear magnetic resonance	Lambert et al., 1996
Lebanon	120-135	Infra red spectrometry	Bandel and Vavra, 1981
		Nuclear magnetic resonance	Lambert et al., 1996
Jordan	120-135	Infra red spectrometry	Bandel and Vavra, 1981
		Nuclear magnetic resonance	Lambert et al., 1996
Switzerland	55-200	Nuclear magnetic resonance	Lambert et al., 1996
Bavaria (Germany)	220-230	Nuclear magnetic resonance	Lambert et al., 1996

Figure 6. Leaf of *Agathis levantensis* in Jordanian sedimentary deposits.

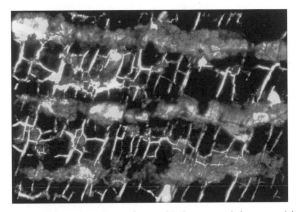

Figure 7. Microscopic examination of *Agathis levantensis* leaves with three parallel veins and outlines of cell walls.

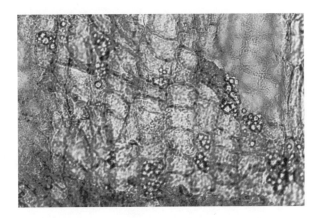

Figure 8. Cellular inclusions in leaf remains of *A. levantensis* .

Table 2. Geographical location of extant kauri pines (*Agathis* spp.) (After Whitmore, 1980 and de Laubenfels, 1988)

Species	Locality	Biome
A. atropurpurea Hyland	Queensland, Australia	Tropical rain forest
A. australis (Lambert) Steud	New Zealand (North Island)	Warm temperate forest
A. borneensis Warburg	Malaysia, Borneo, N. Sumatra	Tropical rain forest
A. celebica (Koord.) Warb.	Malaysia, Borneo, N. Sumatra	Tropical rain forest
A. corbassonii de Laubenfels	New Caledonia	Subtropical rain forest
A. dammara (Lambert) Richard	Philippines, Mollucans, Celebes	Tropical rain forest
A. endertii Meijer Drees	Borneo	Tropical rain forest
A. flavescens Ridley	Malaya	Montane forest
A. kinabaluensis de Laubenfels	Sabah	Upland rain forest
A. labillardiere Warburg	New Guinea, Papua New Guinea	Tropical rain forest
A. lanceolata Lindley ex Warburg	New Caledonia	Subtropical forests
A. lenticula de Laubenfels	Sabah	Montane rain forest
A. macrophylla (Lindley) Masters	New Hebridies	Tropical rain forest, subtropical montane forest
A. microstachya Bailey and White	Australia	Lowland tropical rain forest
A. montana de Laubenfels	New Caledonia	Montane forest
A. moorei (Lindley) Masters	New Caledonia	Tropical lowland forest
A. orbicula de Laubenfels	Sabah, Sarawak	Low montane rain forest
A. ovata (Moore) Warburg	New Caledonia	Lowland rain forest
A. philippinensis Warb.	Philippines, Celebes	Upland rain forest
A. robusta (Moore) Bailey	New Britain, Papua New Guinea, Australia	Tropical and subtropical rain forests
A. silbai de Laubenfels	Espiritu Santo	Lowland rain forests
A. spathulata de Laubenfels	New Guinea	Rain forest

Nature of the Cretaceous Kauri Forest

The ancient Lebanese amber forest probably shared some aspects with the present-day kauri woods of New Zealand and Australia, the major exceptions being that the ancient forest would have contained many less angiosperms and many more reptiles. Evidence suggests that marshy areas dotted the amber forest, which extended to the coast in low-lying deltaic areas (Nissenbaum and Horowitz, 1992). The mature trees of *A. levantensis* probably reached gigantic proportions and ranged from 500 to 1,000 years in age, as does *A. australis* in New Zealand (Figure 10) (Poinar and Poinar, 1994). While the male and female cones are similar to those of other conifers, the flattened broad leaves of kauri pines are quite different from the needle-shaped leaves of what we know as pines today (Figure 1, Plate 2).

Other plant fossils in the Early Cretaceous sandstones of the Zerka Valley, which undoubtedly formed part of the Lebanese kauri forest, included the tree fern, *Weichselia reticulata* (Stokes and Webb) and the cycad *Zamites buchianus* (Etten.) with nearly parallel-sided pinnae. Remains of conifers belonging to the genera *Mesembrioxylon* and *Brachyphyllum* have also been reported, but the family affinities of these is still unclear (Edwards, 1929). Portions of leaves resembling members of the Ginkgoales were recovered from the same beds as *A. levantensis* (Figure 11). These beds also contain pyritized nodules (Figure 12), some of which represent replaced cycadeen cones (Shinaq and Bandel, 1998).

Dark muriform fossil fungal spores resembling *Steganosporium* (Figure 13) and *Alternaria* (Figure 14) have been recovered from crushed

Figure 9. Fossilized female cone scale of *A. levantensis* in sedimentary rock layers in the Zerka valley of Jordan.

Figure 10. Base of an extant kauri pine, *Agathis australis*, in New Zealand.

leaf remains of *A. levantensis*. Both of these genera of Hyphomycetes are reported as plant saprophytes or weak parasites today.

What caused the large resin production of these extinct kauri pines in the Early Cretaceous? Was it a sign of stress resulting from disease or a climate change or was resin production a natural deterrent for large herbivores such as plant-eating dinosaurs? Although there are no published descriptions of dinosaurs from Lebanon, their remains occur on the Arabian Peninsula (Jacobs, 1993). The ranges of several genera of these ancient reptiles reported from the Early Cretaceous of Northern Africa (Weishampel et al., 1990)(Jacobs, 1993) probably extended into the forests and swamps of the Lebanese amber forest.

These dinosaurs (Figure 1) are not the common ones that we hear about today in the popular press and films, but may have included the narrow-headed iguanodontid, *Ouranosaurus*, a 7-meter-long herbivore. Recognized by a distinct skin sail that ran along its entire backbone, this stocky dinosaur presumably reared up on its extended hind legs to gather, with a prehensile tongue, tender leaves into its beak-like mouth. Also most likely grazing in the kauri forest was *Dicraeosaurus*, a 6-ton sauropod reaching 20 meters in length. These, along with other still-undiscovered dinosaurs, would have been prey for the 8-meter-long carnivore, *Carcharodontosaurus*. With flesh-tearing, shark-like teeth 13 centimeters long, this bipedal so- called shark lizard would have been a formidable foe for any creature, even a smaller predator like

Elaphrosaurus. The latter, a 3.5-meter-long ostrich-like dinosaur with a long neck, small head, large eyes, and long, narrow, toothless beak, presumably ran swiftly through the underbrush on its hind legs, snatching birds and lizards off low tree branches with its three-fingered hands (Weishampel et al., 1990).

Some of its victims may have been small pterosaurs which were more abundant than birds at this time (Figure 1). Even snakes, whose occurrence coincided with that of the Lebanese amber forest, may have been prey items. Certainly, representatives of the strange reptilian order known as the Rhynchochephalia were hunted by dinosaurs. If we can assume that the habits of early members of this group were similar to those of the tuatara, the sole survivor of this order, living today on off-shore islands in New Zealand, then the ancient forms, which possessed

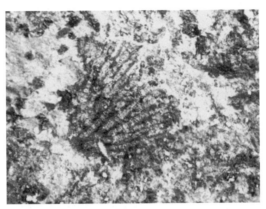

Figure 11. Fossil Ginko-like leaf occuring in same sedimentary deposits as *A. levantensis.*

Figure 12. Iron pyrite crystals from Jordanian sedimentary deposits containing *A. levantensis* leaves.

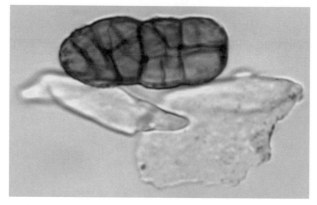

Figure 13. Dark, muriform fossil spore resembling members of the extant genus *Steganosporium* (Moniliales) from macerated leaf of *Agathis levantensis*.

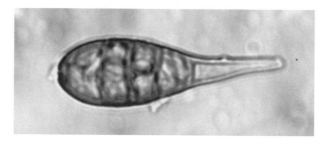

Figure 14. Dark, muriform fossil spore resembling species of the extant genus *Alternaria* (Moniliales) from macerated leaf of *Agathis levantensis* .

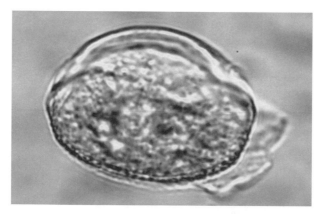

Figure 15. Pollen grain of *Agathis levantensis* from macerated leaf tissue.

both crocodilian and amphibian characters, would have fed on any animal they could catch and swallow (Halliday and Adler, 1986).

Turtles and crocodilians certainly occurred in the ponds and swamps, having replaced the giant amphibians of earlier times, such as *Mastodonsaurus*. Frogs and salamanders, much as we know them today, had already appeared on the evolutionary scene and were undoubtedly represented in the ancient woodland (Halliday and Adler, 1986).

Mammals were certainly represented by various small forms belonging to now-extinct orders. The Multituberculata were widely distributed in the Early Cretaceous and may have occupied niches that are now retained by rodents. They were rat-sized or smaller and probably fed on plants as well as insects and other invertebrates (Lillegraven et al., 1979).

Other now-extinct mammals probably present included tricono-donts, shrew-sized animals with relatively slender, elongate snouts. Like others of their kind, these small creatures had to keep a low profile in the reptile-dominated forest (Figure 1). Similar to triconodonts in size and shape and also likely to be present in the same habitat were the symmetrodonts and eupantotheres. A complete skeleton of the latter, with strong claws and a well developed tail, suggested that it climbed trees and may have dined on birds and lizards (Lillegraven et al., 1979). However, insects and other invertebrates probably composed a major part of the diet for most of these small mammals.

Representatives of blood-sucking flies in Lebanese amber probably fed on all of the vertebrates represented in the forest, including dinosaurs. The scene of dense populations of biting midges surrounding dinosaurs would have provided a striking contrast to the conditions of today's kauri forests. In addition, the sky was probably filled with dragonflies, damselflies, and other insect predators which preyed on these flies.

To date, no evidence of the presence of bees has been found; this is probably a reflection of the low diversity of flowering plants and would have been another distinct difference from today's forests. Modern bees do collect pollen from primitive plants such as cycads, so a possible food source did exist at that early period. The lack of ants, with their many intricate associations with plants and other invertebrates, also would have made this forest quite different from any of the present.

The occurrence of termites, the only social insects known from the Early Cretaceous, suggests that symbiotic associations between insects and cellulose-digesting microbes (probably protozoa) had already become established. With their wood-digesting protozoa, termites provided an environment that supported many other forest animals, both as a direct food source and indirectly through a breakdown of wood products. Cockroaches shared the same micro-habitat as termites in the amber forest, having evolved some 100 million years earlier, when many may have possessed wood-digesting intestinal microbes, such as are retained by a few roaches even today (Cleveland, 1934). The long cockroach lineage prior to their presence in the Lebanese forest is indicated by the existence of two wasps in Lebanese amber that co-evolved with and are now completely dependent on the cockroach. The cockroach wasps of the family Ampulicidae deposit their eggs on their paralyzed victims after dragging them into a shallow grave, while the evaniid wasps lay their eggs inside the cockroach egg cases.

The Shifting Face of Lebanon

Today Lebanon has quite a different character from the land that existed in the Early Cretaceous. Long gone are the equable tropical kauri forests with their dense vegetation of ferns, cycads, and horsetails. The regime today is described as Mediterranean (a subtropical dry summer climate) highly modified by Lebanon's location between a cold-winter continent to the north, a warm inland sea on the west, and the Sahara to the south. This combination results in horizontal air movements which cause unusual temperature fluctuations (advection), and temperature variations at different altitudes. Botanists have divided Lebanon into plant zones from the coastal zone (ranging from sea level to about 1500 feet) to the alpine zone (7500 feet and above). The coastal zone has the most equable climate, with an average daily mean temperature of 13°C in the winter and 29°C in the summer; at approximately 6000 feet elevation in the mountains, the average daily mean temperature varies from 0.1°C in the winter to 18°C in the summer (Larsen, 1974). Human occupancy for the past 5,000 years, resulting in deforestation, overgrazing, and erosion, has greatly modified the natural landscape. Where the ancient kauri forests were located are now orchards and farms with date palms, mulberries, grapes, olives, bananas, oranges, and apples. Of these only the olive is considered native to Lebanon (Hitti, 1962).

Most of the few remaining endemic Lebanese plants and insects occur in the mountains, especially in the area known as the Zone of the Cedar (4500-6000 feet). Aside from the last remaining natural stands of cedar (*Cedrus libani*), rare endemic herbs such as milk vetch (*Astragalus sofarensis*), the Sofar Iris (*Iris sofarana*), and the Romulea iris (*Romulea nivalis*) survive there. Only one endemic coastal plant still remains, a mustard, *Matthiola crassifolia* (Nehmeh, 1978).

Of Lebanese insects, the butterflies have probably been studied more than any other non-agricultural group (Larsen, 1974). As with the plants, the richest butterfly fauna in the country occurs in the Zone of the Cedar. Here reside the only known endemic species such as the Cedar Mountain Blue and Baby Blue of the genus *Lysandra* and the Lebanese hairstreak (*Strymonidia myrtale*).

None of the Lebanese vertebrates are considered endemic, as indicated by many common names such as the Syrian hare, European hedgehog, Persian squirrel, Indian crested porcupine, Asian dormouse, Egyptian fruit bat, etc. The largest mammals reported from Lebanon, such as the Syrian brown bear, leopard, Asiatic wild ass, and Nubian ibex are rarely seen in the wild today (Serhal, 1985). Even more exotic animals appeared in the historic past. During the period of the Crusades there were reports of lions, leopards, and cheetahs within the present-day boundaries of Lebanon (Hitti, 1962). A similar pattern occurs with the resident birds, all of which are common to the Eastern Mediterranean. However, an interesting assortment of migrants pass through Lebanon, including the hoopoe and bee-eater.

The ancient kauri pines have been replaced by yet another noble gymnosperm, the famous cedars of Lebanon (*Cedrus libani*). Although it is questionable whether kauri pines and cedars ever coexisted on the Arabian Peninsula, the fossil record of cedars does extend back to the period when the kauri pine flourished; an extinct fossil cedar known as *Cedrus alaskensis* Arnold was described from Early Cretaceous beds in Alaska (Taylor and Taylor, 1993). Kauri pines and the cedars of Lebanon do have many similarities. Both are large, long-lived, majestic forest trees that reach great heights (40-50 meters). And both bear large resinous female cones, have resinous durable wood, and are much sought after for timber. The story of how Solomon felled large numbers of Lebanese cedars for his Temple parallels that of the Australian and New Zealand shipbuilders felling kauris for masts. Today the remaining kauri pines in the southern hemisphere face the same fate as the cedars as a result of human activity—extinction. Perhaps the closest these two giant tree species have come to each other is at Bqaa Kafra in North Lebanon. At this site, some 6000 feet above the sea, a small deposit of amber from the ancient kauri forest lies buried beneath one of the last stands of the famous cedars of Lebanon.

TYPES OF INCLUSIONS IN LEBANESE AMBER

The various fossils that have been found in Lebanese amber will be discussed under their appropriate taxonomic group, with notes about their probable biology and animal and plant associates. Up to the present a total of some seventy-five species in sixty-three genera, and thirty-four families have been recorded. Four animal phyla are represented (Nematoda, Mollusca, Arthropoda and Vertebrata); insects constitute the largest group, with fifteen orders, twenty-nine families, fifty-six genera, and sixty-nine species (Table 3).

Biological notes and the systematic placement of the inclusions are based on the following works; Borror et al.(1989), Crowson (1981), Daly et al. (1998), Hanson and Gauld (1995), White (1983), Goulet and Huber (1993), and McAlpine (1981; 1987).

Monera

Bacteria are not easy to detect in amber samples but those that occur are usually well preserved. The branching filaments and isolated conidia (Figure 16) typical of *Streptomyces* provide a fleeting glimpse of a colony of Early Cretaceous Actinomycetes.

Figure 16. Filaments and conidia of an actinomycete in Lebanese amber.

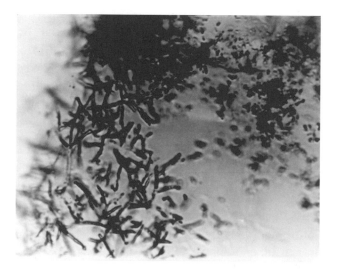

Table 3. Genera, families, and orders of insects described from Lebanese amber (extinct families are underlined, extant genera are in bold type).

Genus	Order (all extant)	Family	Reference
Aphelopus	Hymenoptera	Dryinidae	Olmi, 1998
Archiaustroconops	Diptera	Ceratopogonidae	Szadziewski, 1996; Borkent, 2000
Archiculicoides	Diptera	Ceratopogonidae	Szadziewski, 1996*
Archisciada	Diptera	Sciadoceridae	Grimaldi and Cumming,1999
Atelestites	Diptera	Empididae	Grimaldi and Cumming, 1999
Austroconops	Diptera	Ceratopogonidae	Szadziewski, 1996; Borkent, 2000
Avenaphora	Diptera	Empididae	Grimaldi and Cumming, 1999
Banoberotha	Neuroptera	Berothidae	Whalley, 1980
Bernaea	Hemiptera	Aleyrodidae	Schlee, 1970
Chomeromyia	Diptera	?	Grimaldi and Cumming, 1999
Conovirilus	Ephemeroptera	Leptophlebiidae	McCafferty, 1997
Corethrella	Diptera	Corethrellidae	Szadziewski, 1995
Cretaceomachilis	Archeognatha	Meinertellidae	Sturm and Poinar, 1998
Cretapsychoda	Diptera	Psychodidae	Azar et al., 1999A
Enicocephalinus	Hemiptera	Enicocephalidae	Azar et al., 1999B
Exitelothrips	Thysanoptera	Scudderothripidae	zur Strassen, 1973
Fossileptoconops	Diptera	Ceratopogonidae	Szadziewski, 1996
Glaesoconis	Neuroptera	Coniopterygidae	Whalley, 1980
Heidea	Hemiptera	Aleyrodidae	Schlee, 1970
Incurvariites	Lepidoptera	Incurvariidae	Whalley, 1978
Jezzinothrips	Thysanoptera	Jezzinothripidae	zur Strassen, 1973
Lebambromyia	Diptera	Phoridae	Grimaldi and Cumming, 1999
Lebanaphis	Hemiptera	Tajmyraphididae	Heie and Azar, 2000
Lebania	Diptera	Tipulidae	Podenas et al., 2001
table continues			

*This genus is considered a synonym of *Protoculicoides* by Borkent (2000).

Genus	Order (all extant)	Family	Reference
Lebanoconops	Diptera Ceratopogonidae		Szadziewski, 1996**
Lebanoculicoides	Diptera	Ceratopogonidae	Szadziewski, 1996
Leptoconops	Diptera	Ceratopogonidae	Borkent, 2000, 2001
Libanobythus	Hymenoptera	Scolobythidae	Prentice et al., 1996
Libanochlites	Diptera	Chironomidae	Brundin, 1976
Libanophlebotomus	Diptera	Phlebotomidae	Azar et al, 1999A
Libanopsychoda	Diptera	Psychodidae	Azar et al, 1999A
Libanorhinus	Coleoptera	Nemonychidae	Kuschel and Poinar, 1993
Libanosemidalis	Neuroptera	Coniopterygidae	Azar et al., 2000
Lonchopterites	Diptera	Lonchopteridae	Grimaldi and Cumming, 1999
Lonchoptero-morpha	Diptera	Lonchopteridae	Grimaldi and Cumming, 1999
Megarostrum	Hemiptera	Tajmyraphididae	Heie and Azar, 2000
Mesobolbomyia	Diptera	Rhagionidae	Grimaldi and Cumming, 1999
Mesophlebotomites	Diptera	Phlebotomidae	Azar et al.,1999A
Microphorites	Diptera	Empididae	Hennig,1971
Minyohelea	Diptera	Ceratopogonidae	Szadziewski, 1996; Borkent, 2000
Mundopoides	Hemiptera	Cixiidae	Fennah, 1987
Neocomothrips	Thysanoptera	Neocomothripidae	zur Strassen, 1973
Paleochrysopilus	Diptera	Rhagionidae	Grimaldi and Cumming, 1999
Paleopsychoda	Diptera	Psychodidae	Azar et al. 1999A
Parasabatinca	Lepidoptera	Micropterigidae	Whalley, 1978
Paraberotha	Neuroptera	Berothidae	Whalley, 1980
Phaetempis	Diptera	Empididae	Grimaldi and Cumming, 1999
Phlebotomites	Diptera	Phlebotomidae	Hennig, 1972
Progonothrips	Thysanoptera	Rhetinothripidae	zur Strassen, 1973
Protoculicoides	Diptera	Ceratopogonidae	Szadziewski, 1996; Borkent, 2000
Protopsychoda	Diptera	Psychodidae	Azar et al, 1999A

table continues

**This genus is considered a synonym of *Minyohelea* by Borkent (2000)

Genus	Order (all extant)	Family	Reference
Rhetinothrips	Thysanoptera	Rhetinothripidae	zur Strassen, 1973
Scaphothrips	Thysanoptera	Scaphothripidae	zur Strassen, 1973
Scudderothrips	Thysanoptera	Scudderothripidae	zur Strassen, 1973
Sympycnites	Diptera	Dolichopodidae	Grimaldi and Cumming, 1999
Trichinites	Diptera	Empididae	Hennig, 1970

Fungi

Fungi are ubiquitous and certainly diverse types were present when Lebanese amber was forming. However, because of their delicate nature, most fungi decompose quickly and never become fossilized by standard processes. Fungal spores and other reproductive structures that are blown against the sticky sap may start to germinate. One example shows protoplasm escaping from a gametangium of what is probably a member of the Chytridiales (Chytridiomycetes) (Figure 17); another appears to be a mature zygospore of a member of the Mucorales (Zygomycetes) (Figure 18). Both of these organisms are known to occur on dead plant material today. Some mycelial strands of a saprophytic fungus are shown in Plate 8. Other Levantine amber organisms considered as the reproductive stages of primitive fungi have been described in the genera *Phycomycitis, Peronosporites,* and *Blastocladitis* (Ting and Nissenbaum, 1986).

Plantae

Gymnosperms were the dominant higher plants in the Early Cretaceous and although there are many fossils of this group in Baltic amber, identifiable remains in Lebanese amber are quite rare.

Angiosperms were present but almost certainly did not dominate the scene as they do in the tropics and subtropics today. Pollen attributed to some of the earliest know angiosperms have been found in Early Cretaceous sedimentary deposits in the Levant. These originate from the Hauterivian-Barremian periods which coincide with the age of Lebanese amber (Taylor and Taylor, 1993). Some partial leaves that resemble those of angiosperms appear in Lebanese amber (Plates 9, 10). Other plant parts consisting of branchlets (Plate 11), filaments (Plate 12), hairs (Plate 13), and rootlets (Plate 14) remain unidentified.

Figure 17. Protoplasm emerging from a gametangium of a member of the Chytridiales in Lebanese amber.

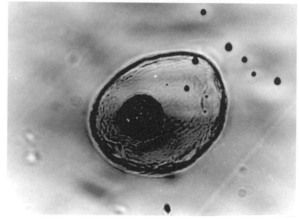

Figure 18. A possible zygospore of one of the Mucorales in Lebanese amber.

The study of pollen and spores in amber is really in its infancy, yet this area could reveal much valuable information on the vegetation present when Lebanese amber was being formed.

Animalia
Nematoda (Nematodes)

Nematodes or roundworms are one of the most plentiful groups of invertebrates on the face of the earth and occur in a wide variety of habitats. Many of the microscopic forms are microbotrophic and feed on bacteria, fungi, and other soil microbes. Representatives of types which occur on the bark of trees, especially in the galleries of wood-boring beetles, are shown in Plate 15. Other nematodes are carried into the resin by their hosts, such as small flies. A mermithid nematode, still within the body of its dipteran host in Lebanese amber, demonstrates the ancient age of this group of now common parasites

(Poinar et al., 1994A). This parasite was originally described in the extant genus *Helidomermis*, based on the understanding that the host was a member of the Ceratopogonidae (Diptera). However, it has now been shown that the host is a member of the Chironomidae (Borkent, 2000) and as a result of this change, the nematode has been transferred to a new genus *Cretaciomermis* (Poinar, 2001).

Mollusca (Snails)

A pupillid land snail in Lebanese amber represents the earliest fossil record of the family Pupillidae. This juvenile specimen is most similar to the described genus *Orcula* Held which ranges from the Paleocene to Recent throughout Europe and the Middle East (Roth et al., 1996). Pupillid snails occur on forest floors and dead wood.

Myriapoda

There is mention of both millipedes and centipedes in Lebanese amber (Borkent, 2000).

Arachnida

Acari (Mites)

Erythraeoidea. The presence of a six-legged larval erythraeid mite (Plate 16) in amber indicates a previous parasitic relationship since extant larvae are known to attack and suck the blood from various arthropods, especially insects. Another erythraeid mite was preserved while sucking blood from its biting midge host (Plates 17, 18). The mouthparts of the mite (near *Leptus* sp.) are still inserted into the body of the fly. After reaching various sites on their hosts (legs, wings, abdomen, thorax), the larval mites puncture the insect's cuticle with their mouthparts and slowly imbibe the hemolymph. When finished feeding, the engorged mites leave the host and molt to the next (nymphal) stage, and assume the role of arthropod predators.

A range of free-living mites occurs in Lebanese amber (Plates 19-22), including a member of the Anystinae (Prostigmata) (Plate 19). Extant examples of these fast-moving mites are predaceous on other mites and small hexapods.

Araneae (Spiders)

Cretaceous spiders are rare and only one, belonging to an extinct genus in the family Oonopidae (Wunderlich and Milki, 2001), has been characterized from Lebanese amber (Plate 23). All fossil members of this family have been found only in amber or copal, probably because they are too small for most other types of preservation. The individual shown here (Plate 23) is 2 mm in length and possesses 6 minute eyes and short legs. This specimen may have been chasing prey such as a small fly or bark louse when it became entrapped in the resin. These spiders do not construct webs but use speed and concealment for ambushing their victims. Some recent members of this family, most of which are tropical or subtropical, can leap backward when disturbed. Modern species spend most of their time under stones or bark or in litter.

Pseudoscorpionida
An undescribed pseudoscorpion has been reported in Lebanese amber (Whalley, 1980).

Hexapoda
A number of Lebanese amber Hexapoda have been noted and/or studied by numerous scientists over the past fifty years. These reports are summarized here along with our observations in order to present as complete a survey of Lebanese amber insects as possible. An early unpublished report by Whalley (1981) provides a valuable starting point for insect fauna from these deposits.

Collembola (Springtails)
Springtails, together with bristletails, are the oldest Hexapoda known, dating back to the Devonian Period, some four hundred million years ago. Collembola occur in marine intertidal, freshwater, and terrestrial habitats and commonly can be seen crawling or jumping in the moist litter zone. Springtails feed mainly on plant material, especially fungal spores and pollen, but are known to suck up nematodes. Many can jump with the aid of an abdominal appendage or furcula which can be suddenly forced down against the substrate. Collembolans are quite primitive in many respects and many species practice what has been termed an indirect method of fertilization. The males deposit stalked spermatophores on the substrate and the females collect the sperm droplet at the top of the stalk. Many springtails possess a collophore or tube on the first abdominal segment which adsorbs water. These wingless hexapods also occur under the bark of trees which explains their presence in amber although some may have been carried into the sticky resin by the wind since they are readily airborne. Those in Lebanese amber, such as the representative of the Arthroplèona shown in Plate 24, await description.

Archeognatha (Bristletails)

Members of this order, along with the springtails, are the most primitive of all Hexapoda and date back to the Early Devonian (Labandeira et al., 1988). These flightless hexapods are often active during the day where they occur in litter and soil. However some species are nocturnal, feeding on algae, lichens, and moss. Like the springtails, bristletails also have an indirect method of fertilization where the males (like the specimen shown in Plates 25 and 26) deposit sperm packets on fine silken threads issuing from the tip of their abdomens. The females pick up the spermatophores and, after fertilization, deposit their eggs in various cracks and crevices. The immatures resemble the adults but lack rod-like projections (styli) on their thoraces. All stages contain styli on their ventral abdominal segments, possess three long tail filaments, and are covered with scales. Eversible vesicles on the ventral surface of their abdomen are used as water absorbing organs.

These primitive hexapods escape their predators by rapid movements or by squeezing into narrow spaces. They can also jump by slapping their abdomens down against the substrate. They prefer concealed habitats such as beneath bark and stones or in dead wood and that is how most probably made contact with the resin.

Lebanese amber contains *Cretaceomachilis libanensis* Sturm and Poinar (1998), the oldest known bristletail of the family Meinertellidae (Plates 25 and 26). This species most closely resembles members of the genus *Machiloides*, which is widely distributed today in both the New and Old Worlds.

Odonata (Dragonflies)

Representatives of this order extend back to the Carboniferous. The larvae are aquatic and prey on small animals in the same habitat. Although the large, predaceous, winged adults rarely occur in amber, a partial specimen in Lebanese amber (a damselfly) was noted by Borkent (2000).

Ephemeroptera (Mayflies)

Lebanese amber contains the earliest known members of the mayfly families Leptophlebiidae (Plate 27) and Baetidae (Plate 28). Mayflies have a long historical record, dating back to the Carboniferous. The larvae (or naiads) have short antennae and mandibulate mouthparts used to ingest mainly plant material. Larval Leptophlebiidae have only two caudal filaments and occur in a variety of aquatic habitats. Members of the Baetidae have three caudal filaments and the larvae usually prefer still water or slow-flowing streams. The males of both families have divided eyes; sometimes the upper portion contains larger facets than the lower. Mayflies are the only known insects which molt as functionally winged individuals. The larvae molt to subimagoes which are sexually immature but fully winged. This stage molts again to form the sexually mature adults. Both the adults and subimagos have only vestigial mouthparts and do not feed. Adults engage in mating flights which often involve large swarms (probably this is how many become entrapped in resin). The leptophlebiid from Lebanese amber, which has been described in the new genus *Conovirilus*, is most closely related to recent forms in South Africa (McCafferty, 1997).

Blattaria (Cockroaches)

Cockroaches are an ancient insect group that has existed since the Carboniferous. In fact, they are the oldest still extant neopterous insects (those with a flight mechanism involving indirect musculature, thus providing the ability to fold the wings over the back at rest).

They prefer warm climates and those that do exist in northern latitudes often invade domiciles in the winter. Their presence in Lebanese amber (Plates 29-31) indicates a warm climate at the time of resin deposition. Cockroaches are scavengers which feed on a variety of plant and animal material. Many live under the bark of trees, which would bring them into contact with resin. It would be interesting to know how many of these roaches possessed intestinal symbionts which would allow them to digest wood, as do both the roach genus *Cryptocercus* and the termites today (Cleveland, 1934). Although they are protected by a shiny hard exoskeleton, deposit their eggs in a tough capsule, and possess a strong chemical defense system, they are attacked by a wide range of predators and parasites, including ampulicid wasps, which also have been recovered from Lebanese amber (see below and Figure 103 on page 203 in Poinar [1992]) and evaniid wasps reported here.

Orthoptera
Grylloidae (Crickets)

Crickets extend back to the Carboniferous, where they probably existed in both terrestrial and arboreal habitats. Crickets are basically plant feeders, although many are also scavengers and some will attack and devour smaller arthropods. Unidentified forms have been reported in Lebanese amber (Whalley, 1981).

Isoptera (Termites)

Schlee (1972) and Spahr (1992) mention the presence of termites in Lebanese amber. These records represent the earliest appearance of this group (there is also one fossil termite reported from the Weald clay of England that is roughly the same age as Lebanese amber), indicating that they evolved before the true ants (Formicidae: Hymenoptera) and are the earliest known social insects. While ant-like wasps, the sphecomyrmids, have been recovered from Canadian and New Jersey Upper Cretaceous deposits, true ants first appear in the early Tertiary (see discussion in Poinar et al., 1999 and Poinar et al., 2000). During the formative years of termite evolution, there were no true ants to shape their defensive behavior. The defense systems employed today by termite soldiers, e.g. slashing with large mandibles or squirting with offensive chemicals, appear to be ineffective against large numbers of ants, which may explain why ants are the most significant predators of termites. What type of interactions occurred between termites and the sphecomyrmids is unknown.

Psocoptera (Bark lice)

Bark lice date back to at least the Jurassic. These insects live in litter and nests, under tree bark, and on vegetation, where they feed on fungi and plant secretions. Bark lice are common amber inclusions since many search for food on the limbs of trees. Undescribed forms occur in Lebanese amber (Whalley, 1981) (Plates 32-35) (Table 6), where they represent about 5 percent of the inclusions.

Hemiptera (Bugs, aphids, scale insects)
Suborder Homoptera
Aleyrodidae (Whiteflies)

Whiteflies are tiny homopterans with powdery wax on their wings and body. They suck juices of various plants and those now found in Lebanese amber could have been feeding on nearby plants as well as on the resin-producing tree. From these deposits were described the extinct genera *Bernaea* and *Heidea* (Schlee, 1970), which are the oldest known whiteflies.

Aphoidea

Aphids, small homopterans that form colonies on a wide variety of plants, seem to prefer cooler climates today, although tropical forms exist. They are found in most if not all Cretaceous amber deposits. Two species belonging to two extinct genera, *Megarostrum* Heie and

Lebanaphis Heie, have been described from Lebanese amber (Heie and Azar, 2000). They were placed in a new subfamily, the Lebanaphidinae Heie in the family Tajmyraphididae Kononova, 1975. This family contains extinct representatives from Siberian, Canadian, and Lebanese amber, representing a span of some 55 million years. Obviously this family was quite successful during the Cretaceous but it then became extinct at the end of the period, probably when the *Agathis* populations were eliminated. The Lebanese forms have very long rostrums, which the above authors note is characteristic of aphids living on trees with rough bark. The host plants of these forms are unknown but *A. levantensis* could be amongst them.

Coccoidea (Scales)

Scale insects comprise a group of small homopterans which are closely associated with their plant hosts. Specimens from Lebanese amber, comprising seven families, four of which are extinct, are being studied by Koteja (2000 and personal correspondence May 6, 2001). One family also has representatives in amber from New Jersey and Canada. It appears that the coccids in the Lower Cretaceous were just as diverse as the recent forms, leading Koteja to believe that coccid radiation must have occurred much earlier than the Cretaceous (Koteja, 2000).

Fulgoroidea (Planthoppers)

Planthoppers are a varied group, ranging widely in form and size. They are especially abundant in warm climates. Many are capable of jumping, especially the brachypterous forms, while those with functional wings generally are strong fliers.

Cixiidae (Cixid planthoppers)

This family, of worldwide distribution, is often regarded as the most primitive of all the planthoppers. Eggs deposited in or adjacent to plant roots hatch into larvae which complete their development underground while the macropterous adults feed on above-ground plant stems (Fennah, 1987). A cixiid planthopper, *Mundopoides aptianus* Fennah (1987) , closely related to the modern genus *Mundopa*, was described from Lebanese amber and represents one of the earliest fossil records of this family.

Other Homoptera

Whalley (1981) mentions the presence of several nymphal leafhoppers (Cicadellidae) and a representative of the family Stenoviciidae (Cercopoidea) in Lebanese amber. The former species may have served as host to the dryinid parasitoid described from these deposits (Olmi, 1998).

Suborder Heteroptera

Members of this suborder are known as true bugs and have the anterior pair of wings more thickened than the posterior pair. Five families are known from Lebanese amber. These include the earliest definite fossils of the Dipsocoridae, Thaumastellidae, and Anthocoridae (Whalley, 1981; Dolling, 1981).

Anthocoridae (Minute pirate bugs)

These are usually quite small, stocky bugs with short heads. Most are predaceous, though many also feed on pollen (Schuh and Slater, 1995). One specialized characteristic in this family is traumatic insemination where the left paramere of the male is modified into an organ that punctures the female's abdomen during copulation. Sperm liberated into the body cavity make their way to the reproductive organs. These bugs have been reported in Lebanese amber by Whalley (1981).

Dipsocoridae

These are small, somewhat flattened and elongate bugs which seek moist conditions associated with damp vegetation (moss) or water sources (in banks, under stones). They are general predators of small arthropods. Members have been reported from Lebanese amber (Whalley, 1981).

Enicocephalidae (Gnat bugs)

These insects look like small assassin bugs (Reduviidae) and tend to be elongate in form; wings may or may not be present. They occur in leaf litter, moss, and loose soil and under bark, mostly in humid tropical and subtropical regions. They are general predators and probably attacked smaller insects on the bark of the resin-producing tree.

The winged adults often form mating swarms which would also account for their occasional presence in amber. A new genus of these bugs, Enicocephalinus, was described from Lebanese amber (Azar et al., 1999B).

Miridae (Plant bugs)

These are small to medium-sized bugs, normally of ovoid shape; however some of the ant-mimics are more elongated. They occur on foliage, flowers, and bark. The food habits are varied. Many suck plant juices but some are predaceous (one group feeds on scale insects on tree bark) and others are associated with fungi (Schuh and Slater, 1995). Because of its long legs and short, stout beak, it is likely that the mirid in Plate 36 was a predator.

Thaumastellidae

These are small, elongate brown bugs which are related to the Lygaeidae and Cydnidae. The adults may be fully winged or brachypterous. Extant forms live under stones and apparently feed on seeds. Forms in Lebanese amber represent the earliest known fossils of this family (Whalley, 1981; Dolling, 1981).

Neuroptera

The Neuroptera or net-winged insects extend back to the Permian. They range from small to large in size and are characterized by long, filiform antennae, mandibulate mouthparts, and two pairs of membranous wings, usually with numerous veins. The larvae, which are predaceous, are mostly terrestrial but a few are aquatic.

Berothidae (Beaded lacewings)

Extant beaded lacewings are rare, slender, small to medium insects that have a worldwide distribution. The antennae are generally shorter than the fore wings and the wings and thorax are often covered with scales. The eggs of known species are stalked and the larvae are predaceous. The larvae of one North American species attack termites and emit a gas from the anus which immobilizes the prey. Two extinct genera have been described from Lebanese amber, *Banoberotha* Whalley (1980) with cursorial forelegs and *Paraberotha* Whalley (1980) with raptorial forelegs.

Coniopterygidae (Dusty wings)

Known to feed on the immature stages of whiteflies and various other small, soft-bodied prey, coniopterygids are some of the smallest of the Neuroptera. A species of *Glaesoconis* (Aleuropteryginae) was described from Lebanese amber by Whalley (1980). This genus was originally described from Siberian Cretaceous amber (Table 6)(Meinander, 1975) and recently has also been documented in New Jersey amber (Grimaldi, 2000). Thus, this was a widely distributed genus in the Cretaceous.

Another extinct genus and species of this family (but in the subfamily Coniopteryginae) found in Lebanese amber is *Libanosemidalis hammanaensis* Azar et al. (2000). This species is considered to belong to a derived, extant lineage, thus indicating that the family is quite ancient.

Myrmeliontidae (Antlions)

One partial specimen of a possible antlion was noted by Whalley (1980) in Lebanese amber. The fossil record of antlions dates back to the Santana Formation of the Upper Aptian (Poinar and Stange, 1996), so if this specimen could be confirmed as a myrmeliontid, it would be the oldest record of the family. Many of the extant larvae of this family build pit fall traps which, as the name implies, catch ants, as well as other prey. However, ants have not yet been reported from these deposits, so if the specimen is an authentic antlion, it would appear that prey other than ants were the victims of any traps it built.

Coleoptera
Suborder Archostemata

This is the most primitive branch of beetles, dating back to the Permian. Thus far, only one family of this suborder has been reported from Lebanese amber.

Micromalthidae

Only a single cosmopolitan species, *Micromalthus debilis* LeConte, occurs in this family today. Adults are small, elongate forms with beaded antennae. The life cycle is unusual for beetles and involves paedogenesis. When the larvae mature and food is plentiful, instead of forming pupae and adults, they produce other living larvae. The adults are apparently only produced under adverse conditions when a dispersal stage is needed. Both larvae and adults occur in logs and under bark. A first instar larva has been noted in Lebanese amber (Whalley, 1981; Crowson, 1981).

Suborder Polyphaga
Boganiidae

Beetles of this family (usually classified in the Cucujoidea) are oblong and somewhat flattened and their antennae lack or contain only a weak club. The elongate larvae are known to feed on pollen of Cycadaceae (considered a primitive trait) or in Australia subsist in the flowers of Eucalyptus. The group is limited to Australia and Africa (Crowson, 1981). A possible member of this family in Lebanese amber was mentioned by Whalley (1981).

Colydiidae (Cylindrical bark beetles)

The earliest known members of cylindrical bark beetles occur in Lebanese amber (Plate 37). These small beetles today can be found under bark, where some feed on a range of fungi and lichens while others are predators, attacking especially other bark-dwelling beetles.

Dermestidae (Scavenger beetles)

The body shape of these small beetles is variable, with the head usually concealed from above and the antennae with a three-segmented club. They occur in a range of habitats, including under bark. Members of this family are known for their habits as scavengers on dried skins and the soft remains of various animals. Some feed on fungi and other plant products. They have been reported from Lebanese amber (Whalley, 1981; Crowson, 1981).

Elateridae (Click beetles)

These beetles have an elongate form with the hind angles of the pronotum usually extending forward. The adults are mostly phytophagous and are found on plant material and under bark. This beetle has the ability to right itself when placed on its back by clicking the segments of its thorax against one another, causing it to spring into the air. The larvae are long and narrow and called wireworms. They are found in the soil, where they feed on the underground portions of plants. The specimen figured here (Plate 38) could have been under bark or may have flown into the sticky resin. Click beetles had previously been reported from Lebanese amber by Crowson (1981).

Mycetophagidae (Hairy fungus beetles)

Representatives of this family are usually found on shelf fungi or under bark where they feed on various stages of fungi. The one shown here (Plate 39) may have emerged from a fungus on the resin-producing tree.

Nemonychidae (Primitive weevils)

The only beetle yet described from Lebanese amber and the only known member of the family Nemonychidae in any amber deposit is *Libanorhinus succinus* Kushel and Poinar (1993). The Nemonychidae is the oldest and most primitive extant family of the Curculionoidea. Characteristic features include stridulary organs on the pronotum and a free labrum not fused to the clypeus. The Lebanese fossil nemonychid belongs to the extinct subfamily Eobelinae which was originally erected for Jurassic representatives recovered in Kazakstan. It appears that this subfamily was restricted to the Northern Hemisphere during the Mesozoic and became extinct sometime in the Cretaceous. Of the twenty-two extant genera of Nemonychidae, all but two develop in the male cones of conifers, especially those of *Araucaria* and *Agathis* but also in members of the Pinaceae and Podocarpaceae (Kushel and Poinar, 1993). Some species develop only on kauri pines in Australia and New Guniea (extant species of *Notomacer* and *Pagomacer* occur on *Agathis robusta* and *A. atropurpurea* in Australia). Adults, as well as larvae, feed on pollen from the host tree. The females deposit their eggs in male cones at the time of anthesis. After completing their development, the larvae drop to the ground and pupate. It is quite likely that *Libanorhinus* developed in the male cones of *A. levantensis*. Tissue removed from this fossil weevil yielded the oldest known samples of DNA (Cano et al., 1992). After the tissue was removed, the weevil was embedded in liquid plastic for protection and preservation.

Salpingidae (Narrow-waisted beetles)

The body of these small beetles varies from oblong to elongate with the head directed forward and the antennae bead-like. Adults can be found in detritus, on flowering plants, or under tree bark, especially of conifers. The larvae are predaceous on wood-boring insects. Mention of this family in Lebanese amber was made by Whalley (1981).

Scarabaeoidea

Records of this superfamily in Lebanese amber were made by Crowson (1981), who depicted a specimen that was difficult to place in any modern family. The adult possessed morphological modifications suggesting that it was termitophilous and could have occurred in termite nests. Crowson (1981) felt that this fossil, possibly belonging to the family Karumiidae, was the earliest evidence of an insect nest inquiline. In their study of this group, Howden and Storey (1992) mention that this specimen also resembles members of the aphodine tribe Corythoderini. An ancient association with termites is possible since these social insects are known from these deposits (Schlee, 1972, and discussion on page 37).

Scolytidae (Bark beetles)

The body of these small beetles is usually cylindrical with short, elbowed antennae and the head concealed from above. Most species live in galleries made between the bark and wood of trees. Eggs are deposited and the larvae develop in this environment. Mention of this family in Lebanese amber by Whalley (1981) would represent the earliest fossil record of bark beetles.

Scydmaenidae (Ant-like stone beetles)

The body of these small beetles is ant-like with long legs and oval wing covers (elytra) that cover the entire abdomen. They occur in concealed, humid habitats such as under bark and stones and in moss. They are predaceous on small invertebrates, especially mites. A record of this family in Lebanese amber by Crowson (1981) represents the earliest occurrence of these beetles.

Trogossitidae (Bark-gnawing beetles)

These small beetles (Plate 40) have a variable body shape ranging from oval to elongate and a three-segmented antennal club. They occur under tree bark where both adults and larvae prey on wood-boring insects. A few feed on fungi and decaying plant material.

Thysanoptera (Thrips)

These small insects are common in moist concealed plant habitats. They are more common in Lebanese than in Canadian and Dominican amber (Table 6)(Plates 41-42). The adults possess small elongate wings, often with a fringe of hairs on their outer margin. The life history of most species is unknown; however many feed on plant material (pollen, spores, nectar, leaves) while some are predatory on smaller mites and springtails. Seven species, *Exitelothrips mesozoicus, Scudderothrips sucinus, Jezzinothrips cretacicus, Neocomothrips hennigianus, Rhetinothrips elegans, Progonothrips horridus* and *Scaphothrips antennatus*, all placed in extinct genera and in four extinct families, have been described from Lebanese amber by zur Strassen (1973).

Trichoptera (Caddis flies)

Although descriptions of Lebanese amber caddis flies have not yet appeared, nine specimens are being studied by Wilfried Wichard (personal correspondence). The presence of caddis flies in these deposits supports other finds of aquatic insects and hints at a range of aquatic habitats in the ancient amber forest.

Lepidoptera

The presence of this order in Lebanese amber is especially interesting in light of present-day relationships between moths and flowering plants. Only a few specimens have been recovered and they all are microlepidoptera, the group most commonly found in amber.

Micropterigidae (Micropterigids)

Whalley (1978) described a primitive moth in the extinct genus *Parasabatinca* in this family. This form belongs to a group now restricted to Australia-New Zealand and Africa and shows some relatedness to the ancient family Agathiphagidae, whose extant representatives breed in the female cones of trees of the genus *Agathis*, suggesting a possible relationship with the Lebanese amber-producing tree. Moths of this family are interesting since they lack a proboscis, but retain mandibles which allow them to feed on spores. The larvae of most extant species of this family feed on moss and lower plants.

Incurvariidae
Larvae of this family often start their life as leaf miners, then as they become too large to remain within the leaf tissue, emerge and build protective, portable cases. The earliest fossil record of this family is represented by specimens from Lebanese amber which Whalley (1978) assigned to the genus *Incurvarites*.

Diptera
True flies of the order Diptera possess two or no wings and are a very successful group today, possessing a range of habits ranging from saprophagous to parasitic on plants and animals. While many cause crop losses, the most economically important members of this order are vertebrate blood suckers that carry some of the most deadly pathogens known to scientists. Of course, the drosophilid fruit flies have contributed much to our understanding of genetics and development. This order dates back to the early Triassic.

Suborder Nematocera
Members of this suborder are considered the most primitive extant flies. They are characterized by many segmented, bead-like antennae and larvae generally with a well-developed head capsule. Most breed in moist or freshwater habitats and many of the adults take blood from either vertebrates or invertebrates.

Chironomidae (True midges)
True midges are small, delicate, mostly non-biting flies whose larvae are often aquatic, feeding on decaying vegetable matter in still or running water (Plates 43, 44). Many also breed in tree holes and rotting wood. *Libanochlites neocomicus* Brundin (1976) is described from Lebanese amber.

Ceratopogonidae (Biting midges)
Biting midges are the most abundant family of insects in Lebanese amber (Plates 45-48; Table 1) and these records represent the earliest fossil records of the family. The larvae live in moist or wet habitats and

many are associated with decaying bark. Adults of both sexes feed on nectar but most females obtain protein by consuming blood from either vertebrates or invertebrates (often from the wing veins or appendages of larger insects—see Plate 64 and discussion under symbiosis) or function as insect predators. Larval habitats are varied and include moist areas around tree holes, in moss and algae on rotting logs, under bark, and in moist soil, including sandy areas along the coast.

A parasitic association involving an erythraeid mite in the process of taking hemolymph from a biting midge is shown in Plates 17 and 18. Some midges in amber still possess well-preserved tissue (note female in Plate 48). The Lebanese amber ceratopogonids have been studied by Szadziewski (1996) and Borkent (2000; 2001). The former author described five new genera, namely *Archiculicoides, Lebanoculicoides, Archiaustroconops, Lebanoconops* and *Fossileptoconops.* Szadziewski (1996) also described three fossil species in the extant genus *Austroconops* Wirth and Lee, which is represented today by a single "living fossil" (*A. mcmillani* Wirth and Lee), a vertebrate blood feeder found only in Western Australia. Borkent (2001) described two extinct species of *Leptoconops* in Lebanese amber. This is the second extant genus of ceratopogonids in these deposits. *Leptoconops* today is composed of some 130 species, which are vigorous biters of mammals, birds, and reptiles(Auezova et al., 1990). They are diurnal feeders which occur in semi-desert as well as moist habitats (in sand on marine or freshwater beaches)(Downes, 1971). The presence of these two extant genera provides indirect evidence of vertebrate groups in the amber forest. While mammals and birds are the preferred vertebrate hosts of biting midges today, some do feed on reptiles, and dinosaurs probably served as a source of nourishment for many of these forms.

Borkent (2000) considered *Lebanoconops* Szadziewski a synonym of *Minohelea* Borkent and *Archiculicoides* Szadziewski a synonym of *Protoculicoides* Boesel (mentioning that members of the latter extinct genus likely fed on vertebrate blood) and described new species of Lebanese amber *Archiculicoides, Archiaustroconops, Austroconops,* and *Minyohelea.* He recognized the primitive character states in *Libanoculicoides mesozoicus* Szadzeiwski by placing it in a separate subfamily, the Lebanoculicoidinae (Borkent, 2000).

Various members of this family today transmit a variety of microorganisms to higher animals (Kettle, 1984). These include arboviruses to mammals, protozoa to birds and mammals, and filarial worms to mammals. Perhaps one day, it will be possible to determine what these biting midges were transmitting to dinosaurs, pterosaurs, and other vertebrates in the Lebanese amber forest.

Cecidiomyidae (Gall midges)

Gall midges are small flies that are recognized mainly by their effects on various plants, both gymnosperm and angiosperm. Swollen, twisted plant tissues often reveal their presence. The larvae may breed in galls on stems and leaves of ferns and some live inside conifer needles. Gall midges are one of the few families of flies that form galls on the above-ground portion of plants. The habits of those in Lebanese amber (Plates 49-50) are unknown.

Corethrellidae

The members of this small family of mosquito-like flies, composed of some sixty-two species today, are all placed in the single extant genus, *Corethrella*. They are mostly tropical and subtropical in distribution. The adults have a short proboscis and C-shaped eyes that curve around the base of the antennae. The feeding habits of females are variable: they take blood from mammals, birds, and frogs (Williams and Erdman, 1968)(McKeever, 1977). Szadziewski (1995) described a member of the extant genus *Corethrella* Coquillett from Lebanese amber. This species, *C. cretacea* Szadziewski, was placed in a new subgenus, *Fossicorethrella* Szadziewski 1995. This family is distributed throughout the tropics and subtropics today and fossils also occur in Dominican and Saxonian amber (Szadziewski, 1995) (Table 4).

Culicidae

A yet unstudied specimen of what may be a mosquito (Azar, personal communication) occurs in Lebanese amber. If confirmed, this find is quite interesting, since it would extend the fossil record of this group of blood-sucking flies back in time some sixty million years. The oldest known fossil mosquito and currently the only described Cretaceous culicid is *Paleoculicis minuta* Poinar et al. (2000) from Canadian Cretaceous amber dated between 75 and 80 million years. A report of a specimen referred to as "the world's oldest mosquito fossil, with mouth parts tough enough to feed on dinosaurs" in New Jersey amber made by David Grimaldi (in Hilts, 1996) was later identified as a biting midge. It would be interesting to learn more about the hosts and vector associations of mosquitoes from the period of Lebanese amber.

Psychodidae (Moth and phlebotomine [or sand] flies)

Lebanese amber contains both types of flies placed in this family: the moth flies, which are mainly saprophytic in habit, and the phlebotomines (Plate 51) which, while saprophytic as larvae, are often vertebrate blood suckers as adults. The latter subfamily is sometimes raised to the rank of family, the Phlebotomidae. In 1972, Hennig described *Phlebotomites brevifilis* and *P. longifilis* from Lebanese amber and, in 1999, Azar et al. described several new genera, *Mesophlebotomites, Libanophlebotomus, Paleopsychoda, Protopsychoda, Libanopsychoda,* and *Cretapsychoda* from Lebanese amber. These are the oldest definite fossils of this group, which is regarded as two separate families (Phlebotomidae and Psychodidae) by Azar et al.(1999A).

Over the years, parasitic organisms such as nematodes and protozoa have become associated with these flies, especially in the Near East (Adler and Theodor, 1929A). Species of the protozoan genus *Leishmania* utilize sandflies as vectors to reach vertebrates, including humans (Lainson and Shaw, 1987). In the Near East, one species in particular, *Leishmania tropica,* is the causal agent of what has been called the oriental sore, now referred to as cutaneous leishmaniasis in humans. In this disease, the protozoans, transmitted by *Phlebotomus papatasi* and other species, cause a cutaneous sore. Gerbils are the natural hosts of this pathogen and humans are infected when they encounter infected sandflies in gerbil-infested areas.

Some species of sandflies feed on reptiles and are capable of transmitting the protozoan *Sauroleishmania* to these cold-blooded vertebrates (Lainson and Shaw, 1987)(Adler and Theodor, 1929B). It is possible that a similar association occurred in the Early Cretaceous between some of the Lebanese amber sandflies and dinosaurs. Perhaps some of the fossil specimens contain protozoan pathogens, the evidence of which would establish an ancient date for a relationship between protozoans and phlebotomine flies.

Scatopsidae (Scavenger flies)

The larvae of scavenger flies are detritivores and feed on both decaying animal and plant matter. Those found in Lebanese amber (Plate 52) may have developed under the bark of the resin-producing tree.

text continues following the plates, on page 57

Plate 1. Gathering Lebanese amber eroded out from sedimentary rocks.

Plate 5. Some amber is opaque, thus masking most fossils.

Plate 2. Branch of a living kauri pine, *Agathis lanceolata*, in Australia. Note the similarity of leaves to those in Figures 5-6.

Plate 6. Circular layers reveal evidence of multiple resin flows flowing around a circular object.

Plate 3. Transparent amber reveals most fossils.

Plate 7. Some amber is deposited in highly carbonized strata.

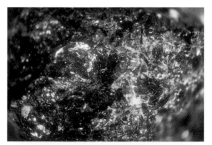

Plate 4. Dark amber often contains unidentifiable plant material.

Plate 8. Mycelial strands probably from a saprophytic fungus developing on the trunk of *A. levantensis* n. sp.

Plate 9. Leaf remains of a possible angiosperm.

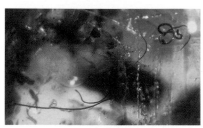

Plate 13. Leaf hairs (trichomes).

Plate 10. Leaf remains of a possible angiosperm.

Plate 14. Rootlets of a possible epiphyte on the amber tree.

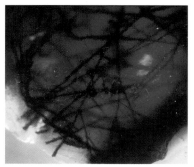

Plate 11. Small branchlets.

Plate 15. Free-living nematodes, probably microbotrophic forms that were feeding under the bark on bacteria or fungi.

Plate 12. Plant filaments.

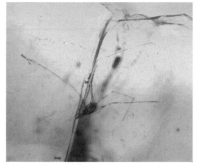

Plate 16. A long-legged erythraeid mite (Acari).

Plate 17. A larval erythraeid mite (*Leptus* sp.; Acari)) feeding on an adult biting midge (Ceratopogonidae: Diptera).

Plate 21. A free-living mite (Acari).

Plate 18. Detail of the mite in Plate 17, with its mouthparts embedded in the body of the fly.

Plate 22. A free-living mite (Acari).

Plate 19. A free-living mite (Acari) belonging to the subfamily Anystinae

Plate 23. An oonopid spider (Oonopidae: Araneae).

Plate 20. A free-living mite (Acari).

Plate 24. An arthropleonid spring-tail (Collembola).

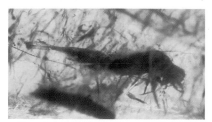

Plate 25. The oldest known member of the bristletail family Meinertellidae (Archeognatha). *Cretaceomachilis lebanensis* Sturm and Poinar 1996.

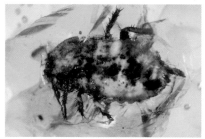

Plate 29. A cockroach (Blattaria).

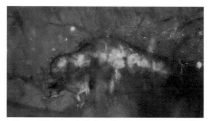

Plate 26. The bristletail, *Cretaceomachilis lebanensis*.

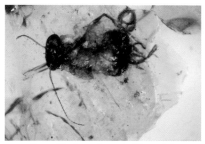

Plate 30. A cockroach (Blattaria).

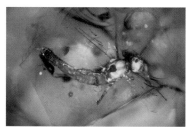

Plate 27. A mayfly, *Conovirilus poinari* McCafferty 1997 (Leptophlebiidae: Ephemeroptera). The earliest fossil representative of this family.

Plate 31. A cockroach (Blattaria).

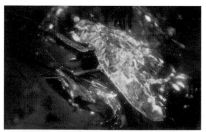

Plate 28. A mayfly of the family Baetidae (Ephemeroptera). The earliest known representative of this family.

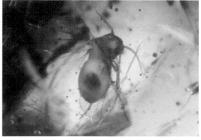

Plate 32. A bark louse (Psocoptera).

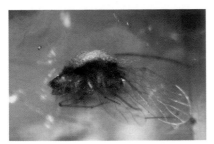

Plate 33. A bark louse (Psocoptera).

Plate 37. A cylindrical bark beetle of the family Colydiidae (Coleoptera).

Plate 34. A bark louse (Psocoptera).

Plate 38. A click beetle of the family Elateridae (Coleoptera).

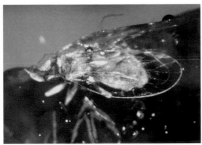

Plate 35. A bark louse (Psocoptera).

Plate 39. A hairy fungus beetle of the family Mycetophagidae (Coleoptera).

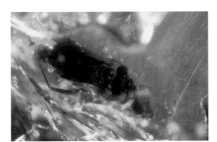

Plate 36. A plant bug (Miridae: Hemiptera), probably with predatory habits.

Plate 40. A bark gnawing beetle of the family Trogossitidae (Coleoptera).

Plate 41. A thrips (Thysanoptera).

Plate 45. A biting midge
(Ceratopogonidae: Diptera).

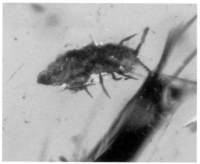

Plate 42. A thrips (Thysanoptera).

Plate 46. A biting midge
(Ceratopogonidae: Diptera).

Plate 43. A
male midge
(Chironomidae:
Diptera).

Plate 47. A biting midge
(Ceratopogonidae: Diptera).

Plate 44. Detail of antennae on male
midge in Plate 47 (Chironomidae: Diptera).

Plate 48. A
biting midge
with exception-
ally preserved
abdominal tissue
(Ceratopogonidae:
Diptera).

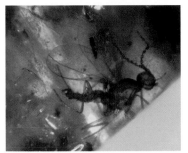

Plate 49. A gall midge
(Cecidiomyidae: Diptera).

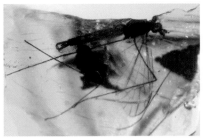

Plate 53. A crane fly of the genus *Lebania*
(Tipulidae: Diptera).

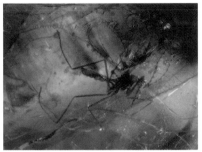

Plate 50. A gall midge
(Cecidiomyidae: Diptera).

Plate 54. A crane fly of the genus *Lebania*
(Tipulidae: Diptera).

Plate 51. A
phlebotomine
fly belonging
to the genus
*Mesophleb-
otomites* Azar
et al, 1999
(Psychodidae:
Diptera).

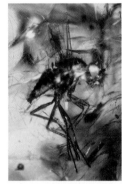

Plate 55 A long-
legged fly
(Dolichopodidae:
Diptera).

Plate 52. A scavenger fly
(Scatopsidae: Diptera).

Plate 56. An unidentified fly (Diptera)

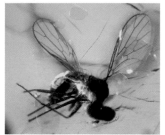

Plate 57. A dance fly close to the genus *Brachystoma* in the subfamily Brachystomatinae (Empididae: Diptera).

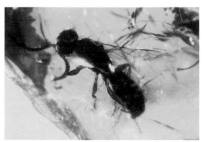

Plate 61. A chrysidid wasp (Chrysididae: Hymenoptera)

Plate 58. A ceraphronid wasp (Ceraphronidae: Hymenoptera).

Plate 62. A member of the Scolobythidae (Hymenoptera), *Libanobythus milkii* Prentice, Poinar and Milki, 1996.

Plate 59. A sac wasp of the extant genus *Aphelopus* (Dryinidae: Hymenoptera). Extant hosts of this genus of wasps are angiosperm-feeding leafhoppers.

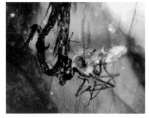

Plate 63. A bethylid wasp (Bethylidae: Hymenoptera)

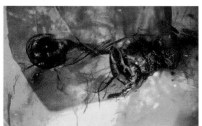

Plate 60. An ensign wasp (Evaniidae: Hymenoptera).

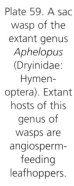

Plate 64. A possible example of symbiosis involving a parasitic biting midge (*Archiaustroconops ceratoformis* Szad.: Ceratopogonidae: Diptera) and the remaining legs of its likely arthropod host.

Sciaridae (Fungus gnats)

Fungus gnats or sciarids are found in moist habitats near the soil, in decaying wood, or under tree bark. The larvae usually occur in damp habitats where they feed on fungi and decomposing plant material. Undescribed members in Lebanese amber represent the oldest known fossils of this family (Whalley, 1981).

Simuliidae (Black flies)

This is a family of small flies that occur around streams. While the majority of adults feed on blood from vertebrates, the larvae are filter feeders in an aquatic environment, usually fast-flowing streams. Mention of this family in Lebanese amber was made by Wichard and Weitschat (1996). This represents the oldest fossil record for this family. While most extant black flies bite warm-blooded hosts, some are attracted to reptiles (Downes, 1971); thus a definite host group cannot be determined for this still unstudied fossil.

Tipulidae (Crane flies)

Crane flies are medium to large insects with long, thin legs, small heads, and usually elongate narrow bodies. The wings are typically large and slender. The larvae live in a variety of habitats, especially wet or damp soil or standing water, where they feed on plant material. The mouth parts of some adults are drawn out into a proboscis as long as or longer than those on mosquitoes. However cranefly probosci are only used to obtain nectar from flowers and not blood from animals. Two new species in the extinct genus *Lebania* have been described from Lebanese amber (Podenas et al., 2001) (Plates 53-54).

Trichoceridae (Trichocerid flies)

Adult trichocerids look like crane flies and prefer dark, damp areas like hollow trees and caverns. The larvae are scavengers and feed both on decaying plant remains and animal refuse. Members of this family have been reported from Lebanese amber (Whalley, 1981).

Other Nematocera

Other Nematocera noted in Lebanese amber include representatives of the Anisopodidae (Borkent, 2000).

Suborder Brachycera

Flies of this suborder tend to be robust forms with less than five (usually three) antennal segments. Most Brachycera larvae and adults are predators of other invertebrates. The larvae, which have a reduced retractile head with vertically moving mouthparts, usually occur in moist concealed habitats.

Asilidae (Robber flies)

Adult robber flies are medium to large predatory insects, with legs often armed with long setae and spines to hold their arthropod victims as they suck out the body contents with their proboscis. Both adult and larval robber flies are predatory on a range of invertebrates. They are rare as fossils so their occurrence in Lebanese amber is of interest (Whalley, 1981).

Chloropidae (Grass flies)

These are small flies found amongst low vegetation. The larval feeding habits are quite varied, ranging from plant material (especially stem borers) to decaying vegetation, fungi, and other arthropods, especially eggs. The larvae of one Australian genus live under the skin of frogs (McAlpine, 1987). Mention of this family in Lebanese amber was made by Whalley (1981).

Dolichopodidae (Long-legged flies)

These small to medium-sized flies with relatively long legs are predaceous in both adult and larval stages. Extant members are often found around trees, which accounts for their relative abundance in amber deposits (Plate 55).

Other interesting flies may belong to this family but require further study (Plate 56).

The oldest definite dolochipodid, *Sympycnites primaevus*, has been described from Lebanese amber (Grimaldi and Cumming, 1999).

Empididae (Dance flies)

These flies are predaceous and usually catch prey while on the wing. They are fairly abundant in Lebanese amber and in Cretaceous amber in general. In 1970 and 1971, Hennig described *Trichinites cretaceus* and *Microphorites extinctus*, respectively, from these deposits and in 1999, Grimaldi and Cumming described *Atelestites senectus*, *Phaetempis lebanensis*, *Microphorites similis*, *M. oculeus*, and *Avenaphora hispida* also from Lebanese amber. The specimen figured here (Plate 57) is a member of the subtribe Brachystomatinae and closely related to the extant genus *Brachystoma*, represented today by four widespread species (McAlpine, 1981).

Lonchopteridae

These are small, slender flies that occur in damp areas near water sources. The larvae are dorsoventrally flattened and possess long filamentous processes on the terminal abdominal and first two thoracic segments. They breed in decaying plant remains. Two species, *Longchopterites prisca* and *Lonchopteromorpha asetocella* , the first fossils of this family , have been described from Lebanese amber (Grimaldi and Cumming, 1999).

Phoridae (Scuttle flies)

The members of this group of flies are relatively small and are characterized by short, sporadic flight patters and frequent running over plant surfaces. The larvae tend to be saprophytic but have wide habits, ranging from commensualism in ant nests to parasitism in both vertebrates and invertebrates. A species of *Lebambromyia* , considered to be a plesiomorphic phorid closely resembling members of the family Ironomyiidae, has been described from Lebanese amber (Grimaldi and Cumming, 1999).

Rhagionidae (Snipe flies)

These medium to large flies usually have an elongated abdomen and longish legs. The adults can be found in damp undergrowth where they prey on other insects. Many wait for prey on tree trunks, which may account for their presence in amber. *Paleochrysophilus hirsutus* and a species of *Mesobolbomyia* have been described from Lebanese amber (Grimaldi and Cumming, 1999).

Sciadoceridae

These small flies are closely related to the phorids. They are attracted to rotting animal and plant material and this is where the larvae occur. *Archisciada lebanensis* has been described in Lebanese amber (Grimaldi and Cumming, 1999)

Hymenoptera

Representatives of this large and diverse order are common in Lebanese amber.

Parasitica

Ceraphronidae

These are small wasps that are parasites or hyperparasites of other wasps that attack aphids and scales. Specimens in Lebanese amber constitute the oldest record of this family (Plate 58).

Chalcidoidea

These small parasitic wasps have not yet been studied in Lebanese amber. Mention of the fairy wasps of the family Mymaridae in these deposits was made by Schlee and Glöckner (1978).

Dryinidae (Sac-wasps)

The Lebanese amber sac wasp shown here (Plate 59), *Aphelopus palaeophoenicius* Olmi, 1998, is the oldest known member of the family. That it is in an extant genus suggests that the dryinids are quite an ancient family. Dryinid larvae are internal parasites of Homoptera, especially leafhoppers (Cicadellidae) and planthoppers (Fulgoroidea). Extant members of the genus *Aphelopus*, which are cosmopolitan and occur on trees, parasitize leafhoppers belonging to the subfamily Typhlocybinae, which feed only on broad-leaf angiosperms (Olmi, 1998)(Hanson and Gauld, 1995). If such host relationships were already established in the Lebanese amber forest, this wasp would provide indirect evidence for the presence of both typhlocybine leafhoppers and angiosperms. Parasitized hosts can usually be recognized by the sac-like extension of their abdomen which contains the developing parasite. While most hosts are killed when the immature wasp emerges, some extant leafhoppers parasitized by *Aphelopus* may continue development to the adult stage after the parasites exit (Hanson and Gauld, 1995). This often results in morphological abnormalities in the adult host ranging from color changes to intersexes. Wasp dispersal is probably dependent on the host since the latter have greater flight abilities than the parasites. Parasite reproduction is by arrhenotoky, where unfertilized eggs form haploid males and fertilized eggs diploid females.

Evaniidae (Ensign wasps)

Ensign wasps are parasites of the egg capsules of cockroaches. The adults are characterized by the presence of a stalked abdomen. The specimen depicted in Plate 60 probably survived on the eggs of the many cockroaches found in these deposits.

Gasteruptiidae

These slender-bodied wasps possess a distinct neck region and the metasome (abdomen) is attached high on the propodeum (thorax). They are cosmopolitan in distribution with most diversity in the tropics. Some larvae are predators on the eggs and larvae of solitary wasps and bees that nest in wood. Mention of this group in Lebanese amber was made by Whalley (1981).

Proctotrupidae

Members of this cosmopolitan family are small, robust wasps that prefer moist habitats. Most are solitary endoparasites of beetle larvae in soil or wood but some parasitize larvae of fungus gnats (Mycetophilidae). Mention of their presence in Lebanese amber was made by Whalley (1981).

Other Parasitica

We have found members of the Scelionidae, small wasps that parasitize insect and spider eggs, also in these deposits.

Aculeates (Higher wasps)

Ampulicidae (Cockroach wasps)

These mainly tropical wasps are considered to be the most primitive of all the Spheciformes (Apoidea). They are rare as fossils, with previous reports only from Baltic amber (Larsson, 1978). Many possess metallic colors, tend to be inconspicuous, and are antlike in behavior (the metasoma is petiolate in many), some actually being ant mimics (Hanson and Gauld, 1995). All known hosts are cockroaches and the wasps' habits of placing the host in a pre-existing cavity and depositing an egg without nest construction or mass provisioning reflects their primitivism. One specimen in Lebanese amber that was mistaken for a possible ant has been identified as belonging to the subfamily Ampulicinae and measures only 1.5 mm in body length (Prentice, 1994). Another larger specimen is also mentioned by Prentice (1993). These represent the earliest fossil records of this family.

Chrysididae (Chrysidids)

The larvae of this family are parasitic on other insects, especially wasps and bees. The adults are often brightly metallic and heavily armored for protection from the adults of the larvae they are parasitizing. Those from Lebanese amber (Plate 61) have not yet been studied.

Formicidae (True ants)

There have been reports or suggestions of ants in Lebanese amber (Schlee and Glöckner, 1978) (Poinar, 1992); however thus far, no true ants of the family Formicidae or even wasp-ants of the families Sphecomyrmidae or Armaniidae, which are known from the Late Cretaceous, have been confirmed in these deposits. A specimen figured on page 103 in Poinar (1992), which was thought by some to be an ant, was later identified as an ampulicid wasp (Prentice, 1994) (see also under section on Ampulicidae above). Were populations of ants and wasp-ants so sparse that they rarely became trapped in resin, was their range restricted, or had they not yet appeared on the evolutionary scene? Thus far, all reports of Early Cretaceous ants, including *Cariridris bipetiolata* (Brandão et al., 1989) from the Santana Formation in Brazil and *Cretoformica explicata* Jell and Duncan (1986) from the Koonwarra beds in Victoria, Australia, are now not considered to be true ants (Baroni Urbani, personal communication).

The presence of ants in an ecosystem can often be determined indirectly by discovering other insects, such as ant mimics, certain homopterans and lepidopterans which have morphological modifications indicating relationships with ants, insects known to live in ants' nests, and even specific predators (Poinar and Poinar, 1999). However, thus far, such insects have not been reported in Lebanese amber, in contrast to the termitophilous beetle (see above under Scarabaeoidea) which indicated the presence of termites in these deposits (see above under Isoptera). It would appear to be much more likely that wasp-ants or sphecomyrmids would be found in Lebanese amber. Unfortunately we know almost nothing about their biology other than the fact that they occurred on trees. There is currently dissension as to whether the sphecomyrmids are true ants or simply a separate family of the Formicoidea. Undisputed true ants of the family Formicidae appear only at the beginning of the Tertiary, (Poinar et al., 1999, 2000).

Scolobythidae (Scolobythids)

Members of this family, which is quite small today, are parasites of wood-boring insects, especially larvae of long-horned beetles. *Libanobythus milkii* Prentice, Poinar and Milki (1986) from Lebanese amber is the oldest known scolobythid (Plate 62).

Other aculeate wasps occur in Lebanese amber but few have been studied in detail. The specimen depicted in Plate 63 has similarities to members of the family Bethylidae, which are exclusively parasites of larvae of Coleoptera and Lepidoptera.

Vertebrata

A beautifully preserved feather has been characterized from Lebanese amber (Schlee, 1973)(Schlee and Glöckner, 1978). This is the oldest known complete fossil feather with tissue remains. Older impression fossils no longer contain organic matter. Certain characters of the feather suggested that it might have come from an aquatic bird and Roxie

Laybourne, a feather expert at the Smithsonian Institution (personal communication), suggested a grebe as a possible candidate.

This find is quite significant since it fits the evolutionary scenario presented by Unwin (1988) regarding the habits of Early Cretaceous birds. At this period, pterosaurs, including members of the genus *Pterodactylus* with a size range from sparrows to gulls, were much more abundant than birds. However pterosaurs were not well developed for land movement and apparently were unable to exploit wet, marshy habitats. Thus these sub-aquatic environments were available to the early birds. Grebes are completely aquatic diving birds that prefer lakes and ponds with much vegetation. Although their fossil record presently extends back only to the Miocene (Unwin, 1988), their lineage may have existed much earlier. The Hesperornithiformes, which do date back to the Early Cretaceous, consist of flightless swimmers and divers, considered by some to be related to present-day grebes and loons. While this similarity could be due to convergence, the question remains whether the Lebanese amber feather is that of a member of the Podicipedidae (grebe family) or whether its grebe-like characters reflect convergence and it belongs to one of the early groups like the Hesperornithiformes. No other vertebrate remains have been reported thus far from Lebanese amber.

Discussion

The fossils reported here are quite significant in illustrating patterns of origination, evolution, distribution, and extinction. The fifty-six genera reported (Table 3) denote the earliest known representatives of an insect ecosystem in amber and contain the first records of many families. Although studies on Lebanese amber insects are in the early stages, some generalities can be made based on the inclusions discussed above.

Symbiotic Associations

It is rare to discover evidence of symbiosis in the fossil record. Due to the quick death of organisms falling into sticky resin, amber is the best medium for revealing various types of past relationships. In Lebanese amber, external or ectoparasitism is represented by parasitic mites attached to the body of their biting midge hosts, as has been discussed earlier (see also Poinar et al., 1994B). Another image shows a biting midge (*Archiaustroconops ceratoformis* Szad.) adjacent to the leg of a large insect (Plate 64). The females of many species of extant biting midges feed on the blood of invertebrates, especially insects (Wirth, 1956). Borkent (2000) concluded that biting midges that feed on arthropods and have enlarged claws fed on small insects (one to three times their own size), while those with small claws fed on larger insects or vertebrates. The midge figured here possesses short, simple, equal claws, which suggests that it fed on large insects. It is possible that this ceratopogonid was feeding on the larger arthropod when both became entrapped in resin.

Internal parasitism is represented by a mermithid nematode, *Cretacimermis libani* (new generic name, see Poinar [2001]) that was still in the abdomen of its chironomid host (Poinar et al., 1994A). The above cases represent the earliest records of external and internal parasites of insects.

Insect Diversification and Distribution

One of the major factors causing insect diversification during the Early Cretaceous was the emergence and subsequent spread of the angiosperms. As various new genera and species of flowering plants appeared, insects evolved new genera and species to exploit these novel food sources. Parasites and predators in turn adapted to these new hosts.

Insect diversification in the Tertiary and Cretaceous can best be demonstrated with genera. The best-studied insect group in Lebanese amber (as well as in many other Cretaceous amber deposits) are the biting midges or Ceratopogonidae. They are well represented in amber because their behavior is conducive to landing on or being blow up against tree branches and becoming entrapped in resin. Thus the distribution of biting midge genera since the Early Cretaceous can be obtained from records in amber. Some genera reported from Lebanese amber occur in other amber deposits (Table 4) and provide data on diversification and distribution.

For example, *Leptoconops*, found in Lebanese amber, also occurs in Siberian, Sakhalin, French, New Jersey, Canadian, Hungarian, and Baltic amber (and is still extant) (Table 4). Its cosmopolitan distribution since the Early Cretaceous indicates a successful host association that has continued for at least some 135 million years. Other ceratopogonid genera show extensive ranges in the Cretaceous but have not been reported in the Tertiary (Table 4). One genus of dusty wings, *Glaesoconis*,

**Table 4. Lebanese amber insect genera reported from other amber deposits.
Data taken from Szadziewski (1996) and Borkent (1995, 2000) unless otherwise noted. Extant genera are in bold type.**

Genera (Family)	Amber source[1] with approximate ages									
	Spanish 100-113	French 91-97	Siberian 80-105	Can. 77-80	Sak. 50-55	N. J. 65-97	Austr. 127-130	Baltic 40	Dom. 15-40	Sax. ?20-30
Aphelopus (Dryinidae)										
Archiaustroconops (Ceratopogonidae)	+								+[2]	
Austroconops (Ceratopogonidae)		+	+							
Corethrella (Corethrellidae)									+[3]	+[3]
Lebanoconops (Ceratopogonidae)			+	+				+		
Glaesoconis (Coniopterygidae)			+[4]			+[5]				
Leptoconops (Ceratopogonidae)[6]		+	+	+	+	+		+		
Minyohelea (Ceratopogonidae)				+			+			
Protoculicoides (Ceratopogonidae)	+			+						

[1]Sources, Can.= Canadian; Sak.= Sakalin; N.J.= New Jersey; Aust.= Austrian; Dom. = Dominican; Sax.= Saxonian (Bitterfeld)

[2]Olmi, 1997; [3]Szadziewski, 1995; [4]Mienander, 1975; [5]Grimaldi, 2000; [6]This genus is also represented in Hungarian amber (80-90 my) (Borkent, 2001).

also had a fairly extensive distribution in the Cretaceous. Other Lebanese amber genera, such as the dryinid *Aphelopus*, which occurs in Dominican amber (Table 4), probably had similar wide distributional patterns, but has not yet been recovered from other amber sources.

Extinctions: Generic Lineages

What can Lebanese amber insects tell us about extinction events? Since all of the insects in Lebanese amber (and from other Cretaceous deposits) belong to extant orders, it is obvious that these groups evolved much earlier (Labandiera et al., 1988). While extinctions of Lebanese amber insects have occurred at the species, genus, and family level, it is impossible to state when these occurred. Since there were many significant abiotic changes during the past 130 million years, extinctions of specific Lebanese amber insects could have taken place at various times during the Cretaceous or Tertiary.

Consensus is that average insect species survive for about two to three million years (the late Frank Carpenter, personal correspondence) although some appear to survive for considerably longer. This is not radically different from estimates for species of other groups. Mammalian species last for about one and a half million years (Savage, 1988), while blastoid echinoderm species survive for some five million years (Paul, 1988). Survival depends on genetic variability and the higher the systematic rank, the greater the variability in that lineage. Genera are more likely to survive than species, families more likely than genera and so on (except in cases of monotypic genera, etc.). It is curious that while much has been written on extinction, relatively little attention has been given to survival tactics over time.

Unfortunately, species as distinct entities in insect paleontology have little significance because comparing fossils with their present-day

Table 5. Characteristics of extant insect genera in Lebanese amber

Genus	Species worldwide	Host	Distribution
Aphelopus[1]	73	leafhoppers	Cosmopolitan
Austroconops[2]	1	mammals	W. Australia
Corethrella[3]	62	mammals, birds, frogs	Cosmopolitan
Leptoconops[4]	122	mammals, birds, lizards	Cosmopolitan

[1]Olmi, personal communication; [2]Szadziewski, 1996; [3]Szadziewski, 1995; [4]Borkent, 1995, 2001.

descendants is nearly impossible. Fine morphological details that define modern insect species (especially those associated with reproduction) are frequently not visible in fossils. There are exceptions in well-preserved amber specimens, but it is often a challenge even to identify extant species when their intraspecific variation is poorly known. Therefore reliable records for insect lineages from the Tertiary and Mesozoic are based on genera and higher categories, although with impression fossils even genera can be difficult to equate with modern taxa.

There are other problems in equating fossils with extant genera, even when diagnostic characters are evident. Could morphological similarities be the result of convergent evolution (homoplasy)? Are the established generic characters too broad and do they in fact include several genera within their confines? And if a particular character is very conserved, could two or more genera have the same morphotype? Finally, it becomes important that identifications be made by those with a familiarity with the extant fauna of that group.

Keeping in mind the above limitations, four genera of Lebanese amber insects are considered to be still extant (Table 5): the leafhopper parasite, *Aphelopus* (Dryinidae:Hymenoptera), the midge *Corethrella* (Corthrellidae: Diptera) and the biting midges *Austroconops* and *Leptoconops* (Ceratopogonidae: Diptera).

These four genera represent the oldest generic lineages known for any terrestrial multicellular animal, having survived the catastrophic events at the end of the Cretaceous and various environmental changes throughout the Tertiary. Longevity of genera in the fossil record is rarely tabulated, but a few mammalian genera are know to have existed between thirty and forty million years (Savage, 1988) and genera of blastoid echinoderms normally survive for about ten million years (Paul, 1988).

To what do these insect lineages owe their success? Is it due to a wide base of genetic diversity that allows them to adjust to differing food sources and climatic changes? Developing nutritional relationships with "successful" hosts is an essential prerequisite for any parasite, including the present four genera, three of which are vertebrate ectoparasites (blood suckers) and one a leafhopper endoparasitoid. Extant *Corethrella* feed on several vertebrate groups, including frogs (Szadziewski, 1995). The single known species of *Austroconops*, a "living fossil," so to speak, feeds on mammalian blood in Western Australia

(Borkent et al., 1987) and members of the genus *Leptoconops* are cosmopolitan vertebrate blood feeders (Table 5). Extant *Aphelopus* parasitize a widely distributed subfamily (Typhlocybinae) of angiosperm-feeding leafhoppers (Cicadellidae) (Olmi, 1998, personal correspondence).

Three of these genera have a wide distribution today and apparently all four were widespread in the Early Cretaceous (Tables 4, 5). Only the *Austroconops* lineage, which is limited to a single species restricted to Western Australia, appears to be on the verge of extinction. The present and past distribution of these genera suggests that at least three are eurytopic (able to adapt to changing conditions) in regards to their climatic tolerance and feeding habits, while *Austroconops* may be stenotopic (adapted to a particular climatic regime). The climatic conditions under which a lineage evolved probably determine whether it is eurytopic or stenotopic. Forms that evolved under temperate conditions can often survive in the subtropics and tropics and are considered eurytopic. Lines that evolved under tropical conditions rarely are able to survive outside of this climatic regime and can be regarded as stenotopic. This degree of flexibility in a lineage based on its origin has been called "genetic confinement" (Poinar, 1992). Since the Lebanese amber environment was tropical-subtropical, the three extant genera with cosmopolitan distributions (Table 5) probably evolved under temperate conditions and were already widely dispersed before the Early Cretaceous.

It is interesting that *Austroconops* only survives in the southern hemisphere today, a pattern also shown by many extinct groups in Dominican amber (Poinar and Poinar, 1999), as well as for *Agathis*, the tree genus responsible for the formation of Lebanese amber and other amber deposits in the northern hemisphere (Table 1). This phenomenon of southern hemisphere survival, or southern hemisphere endemism with the absence of living lineages in the northern hemisphere (but with fossil records) has been reviewed by Newman (1996).

In Lebanese amber, all species and 93%of the fifty-six genera thus far described are extinct. In contrast, of 423 genera reported from Dominican amber (fifteen to forty million years ago) (Poinar and Poinar, 1999), some 92% are still extant. The majority of families (~ 80%) in Lebanese amber are extant and all are extant in Dominican amber (only two subfamilies are extinct). The extinction of insect genera over time shows a correlation with the age of the deposit (Poinar, 1992); however the important factors over time are abiotic changes (especially temperature and moisture). So extinction events are not simply the result of age (time) but are correlated to the amount of environmental change.

The great majority of fossiliferous amber deposits occurred under subtropical and tropical conditions, the climate under which the major resin-producing trees thrived. The insects that inhabited amber forests that experienced extensive climatic shifts following resin production would be expected to have more lineage extinctions than those that evolved under more stable conditions. In the well-studied Canadian deposits dated at 79 million years, only two of the reported seventy-eight genera are considered extant today, *Dryinus canadensis* Ponomarenko (1981) and *Leptoconops primaevus* Borkent (1995)(Pike, 1994). While the climate in both Lebanon and Canada was tropical-subtropical when the amber was being formed, the latter locality experienced a much longer period of cool temperate climate than the former and consequently Canadian amber would be expected to have fewer lineages still existing, even though it is younger in age. Similarly, the high rate of extant insect genera in Dominican (Poinar and Poinar, 1999) and Mexican (Poinar, 1992) amber could be due to the relatively moderate climatic fluctuations that have occurred in that part of the world since their formation (Poinar and Poinar, 1999).

This reasoning leads to the question: could it be possible for genera to exist indefinitely if environmental conditions remained constant? This has been discussed in relation to other fossils and evoked the "Red Queen" and "Stationary Model" hypotheses (Benton, 1990). While it is reasonable to assume that a constant rate of genetic variation occurs, as implied in the "Red Queen Hypothesis" (Van Valen, 1973), it is also logical that selection mainly occurs during periods of environmental change, as implied in the "Stationary Model" (Stenseth and Maynard Smith, 1984). This would explain periods of stasis alternating with those of rapid evolutionary change in the fossil record, which has been described as "Punctuated Equilibria" (Eldredge, 1985). The periods of evolutionary change would indicate adaptive radiation following an environmental change. Lineages that can tolerate the greatest degree of environmental change (eurytopic) or can produce the highest degree of variability through genetic recombination, jumping genes, and mutation would have the best chance of survival. Benton (1990) brings up another interesting aspect of evolution. Will later members of a lineage be competitively superior to earlier ones? While this is difficult to determine, there is no indication of terminal lineage superiority in the present work. In fact, the Law of Constant Extinction (which states that the probability of extinction within any group remains constant through time) would seem to apply here.

While discussion here centers around extinction and evolutionary change as a result of abiotic factors, it should be noted that biotic factors could produce the same effect. Unfortunately the fossil record of parasitic associations, disease-causing organisms, appearance of competitors, loss of food sources, and other biotic factors is incomplete, so that it is difficult to present evidence of extinctions based on these causes. Most paleontological data of extinctions are based on the fossil record of marine invertebrates and, even today, our knowledge of parasites and pathogens in these groups is quite limited. Thus any effects of biotic agents could go unnoticed (Boucot, 1990).

Is it a coincidence that the four extant insect genera in Lebanese amber and the two in Canadian amber all have a parasitic life style? Do lineages of parasitic insects associated with a successful host group have a greater potential for survival than scavengers and herbivores?

All of the extinct families of Lebanese amber insects that belong to the orders Hemiptera and Thysanoptera include plant-feeders. The aphids, coccids, and whiteflies were probably highly host specific, possibly feeding on *A. levantensis*, and it is likely that these groups disappeared at the family level with the demise of the amber forest. Today's kauri thrip, *Oxythrips agathidis* Mor. (Thripidae), that feeds on the leaves of *Agathis robusta* in Australia (Poinar, 1990), provides evidence that some of the Lebanese amber thrips could have fed on *Agathis levantensis*.

Insect-plant Associations

Insect-plant associations in the fossil record can be construed from damaged plant tissue, insect gut contents, and fecal pellet composition (Labandeira and Sepokoski, 1993). Plant damage that could have been caused by insects has not yet been reported from Lebanese amber, although it is known from other amber deposits, including Dominican (Poinar and Poinar, 1999). Gut contents are rarely discernible in insects preserved in amber and, although fecal pellets are common in amber deposits, it is often difficult to connect the pellet with a specific insect group and then to identify the plant or animal species that contributed to the pellet. Indirect evidence of insect-plant associations is based on the presence of phytophagous insects in the amber deposit and use of the law of behavioral fixity in the fossil record (Boucot, 1990). This dictum states that once a family of organisms has become established, their behavior tends to remain constant over time. In other words, the presence of whiteflies, aphids, and coccids in Lebanese amber implies plant-insect associations, since all extant members of these groups are phytophagous.

Insect Population Structure Over Time

While abiotic and biotic changes since the Cretaceous, including the extinction events at the Cretaceous-Tertiary (K-T) border, did not appear to extirpate any insect order, changes in family and genus configuration definitely occurred.

A comparison of the percentage of various arthropod groups found in three separate amber deposits (Lebanese, Canadian and Dominican), all separated by approximately 50 to 60 million years and ranging from 130 to135 to 15 to 45 million years is shown in Table 6. While representatives of the same orders generally occur in each deposit, the taxonomic composition of orders and families differs greatly.

Table 6. Comparison of common arthropod groups in amber from Lebanon, Canada (Alberta), and the Dominican Republic (measured in percentage of total number present).

Group	Lebanese (130-135 mya) (present study)	Canadian (79 mya) (Pike, 1995)	Dominican Republic (15-45 mya) (Poinar and Poinar, 1999)
Arachnida	8	13	3
Psocoptera	5	2	6
Collembola	<1	<1	<1
Archeognatha	<1	<1	<1
Ephemeroptera	1	—	—
Blattaria	4	<1	<1
Orthoptera	1	—	<1
Thysanoptera	3	<1	<1
Homoptera	5	28	3
Heteroptera	<1	<1	<1
Coleoptera	3	9	10
Diptera	51	25	34
Hymenoptera	10	12	37

The greatest number of arthropods in Lebanese amber are flies (Diptera), while in Canadian amber it is aphids (Homoptera) and in Dominican amber, ants (Hymenoptera) (Table 6). The absence of true ants (Formicidae) in Lebanese and Canadian amber must have made the forest unique, since today it is hard to imagine any ecosystem without ants. The dominance of ants in the Tertiary is evident from their high numbers in Dominican amber. These differences reflect evolutionary changes over millions of years and also give clues to the type of ecosystem that predominated.

An estimate of insect composition in the three above-mentioned amber deposits showed some general trends. The categories examined were detritivores—omnivores, plant feeders, predators, parasitoids, and ectoparasites. In comparison to the total population of insects, ectoparasites comprised the highest percentage in Lebanese amber (15%), declined slightly in Canadian amber (10%), and represented only 4% of the total in Dominican amber. Virtually all ectoparasites in Lebanese amber are biting midges (Ceratopogonidae) and phlebotomines (Phlebotomidae), especially the former. Such large numbers suggest an abundance of both invertebrate and vertebrate hosts.

Another population change noted was an increase in predators, from 12% in Lebanese amber to 22% in Dominican. This may be attributed to the larger number of predators attacking phytophagous insects as well as many specialized predators that evolved in association with ants and bees. While plant-feeding insects increased from 16% in Lebanese amber to 30% in Dominican amber, Canadian amber had the highest percentage (45%), with its abundance of aphids. Thus far aphids are represented by only one extinct family in Lebanese amber and only two described species in Dominican amber (there are no aphids described from Mexican amber, but quite an assortment in Baltic amber). In Dominican amber, the niches for aphids are occupied by planthoppers, leafhoppers, and treehoppers. On the basis of their present distribution and the fossil record, it would appear that aphids evolved in the northern hemisphere under a warm-temperate to subtropical climate, where they first diversified on various gymnosperms. Relatively few aphid lineages were able to survive in the tropics, meeting competition from other homopterans that had already evolved under those conditions. In Dominican amber, the most abundant phytophagous family is the Cecidiomyidae (gall midges), a group that certainly radiated with the diversification of the angiosperms, although some are detritivores and a few are predators.

Figure 19. A fossil fish (*Aipichthys velifer* Woodward) from the Upper Cretaceous deposits of Lebanon.

No significant change occurred in the percentages of parasitoids (10% in Lebanese, 9% in Canadian, and 7% in Dominican amber); however some differences were noted in populations of detritivores/omnivores (42% in Lebanese, 25% in Canadian, and 35% in Dominican amber) over the 100-million-year period.

Comparison of Amber Taxa with Lebanese Fish Fossils

Lebanon has another rich supply of fossils, namely marine fish and invertebrates deposited in fine limestone beds when the land was submerged under the Tethyan Seaway some 95 million years ago during the Late Cretaceous (Brown and Lomolino, 1998). During this period, sediments from both northern and southern continental masses accumulated at the bottom of the sea to form the limestone rocks which form the bulk of Lebanon today.

This Tethyan Seaway stretched around the world in a circum-equatorial pathway and provided a tropical environment for a variety of marine fauna. These warm currents were eventually interrupted by drifting continental landmasses and replaced by cold circum-antarctic currents, causing regional extinctions of the thermally intolerant marine life.

Later, during the Tertiary Period, mountain-forming forces and folding moved the marine beds well up above sea level and into the ranges known today as the Lebanon and Anti Lebanon. The fossils, which occur in light buff sandstone as well as marine limestone (Figure 19) (Graffham, 2000), are now being unearthed at Sahel Alma, Hakel, and Hadjula. They testify to the presence of this tropical seaway by revealing a range of organisms that became extinct at the end of the

Table 7 . Genera (all extinct) of fish fossils from Upper Cretaceous, Lebanon, with family and ordinal status (From Frickhinger, 1991).

Genera	Order	Family
Acrogaster	R	R
Aipichthys (Fig. 19)	R	R
Anguillavus	R	R
Apateopholis	R	R
Aphanepygus	E	E
Centrophoriodes	R	R
Charitosomus	R	R
Cheirothrix	R	E
Coccodus	E	E
Cryptoberyx	R	E
Ctenocephalichthys	R	R
Ctenothrissa	E	E
Cyclobatis	R	E
Davichthys	R	R
Dercetis	R	E
Dinopteryx	R	R
Enchelion	R	R
Eubiodectes	E	E
Eurypholis	R	E
Exocoetoides	R	E
Gaudryella	R	E
Gharbouria	R	E
Hajulia	R	R
Hakelia	R	R
Halec	R	E
Hemisaurida	R	E
Heterothrissa	E	E
Hexanchus	R	R
Humilichthys	E	E
Ichthyoceros	E	E
Lebonichthys	R	R
Libanoberyx	R	R
Lissoberyx	R	R
Macropomoides	R	R
Mesiteia	R	R

R = Recent (taxon still exists); E = Extinct

Genera	Order	Family
Micropristis	R	E
Nematonotus	R	R
Omosoma	R	R
Opistopteryx	R	R
Ornategulum	R	R
Osmeroides	R	E
Palaeobalistum	E	E
Pararaja	R	R
Paraspinus	R	R
Paratriakis	R	R
Pateroperca	R	R
Pattersonichthys	E	E
Petalopteryx	E	E
Pharmacichthys	E	E
Phoenicolepis	E	E
Phylactocephalus	R	E
Plectocretacicus	R	E
Plesioberyx	R	E
Prionolepis	R	E
Pronotacanthus	R	R
Protobrama	E	E
Pseudoberyx	R	R
Pycnosterinx	R	R
Pycnosteroides	R	E
Rhinobatos	R	R
Rhombopterygia	R	R
Rhynchodercetis	R	E
Scapanorhynchus	R	R
Sclerorhynchus	R	E
Scombroclupea	R	R
Scyliorhinus	R	R
Spaniodon	R	R
Stichocentrus	R	E
Stichopteryx	R	R
Telepholis	R	E
Urenchelys	R	E

R = Recent (taxon still exists); E = Extinct

Cretaceous or beginning of the Tertiary. Included are mollusks, shrimps, crabs, and lobsters, but the most famous and well studied are the fish fossils, including representatives of guitar fish, sharks, rays, bowfins, and teleosts (Table 7). These fossils were first reported in 1248 in the biography of a Crusader (Hitti, 1962).

Although the fish fossils are younger than the amber fossils (they date from the Cenomanian of the Middle Cretaceous [~95 million years ago] , as compared to the Hauterivian from the Lower Cretaceous [~130 mya], it is interesting to compare the extinction rate of fish genera, families, and orders with those of Lebanese amber insects (Tables 4 and 7). Of the seventy-one genera of Lebanese Cretaceous fossil fish listed by Frickhinger (1991), all are now extinct, as well as 49% of the fossil fish families. In addition, 18% of the fish fossils have been placed in extinct orders. In Lebanese amber, 7% of the genera, 80% of the families, and all of the orders are extant. There is clearly a higher extinction rate in the fish taxa over the past 95 million years than in the amber insect fauna over 130 million years. Although we are comparing vertebrates with invertebrates, this does support earlier discussions suggesting that time itself is not as important as abiotic factors in determining the duration of lineages.

CULTURAL ASPECTS OF LEBANESE AMBER

This section deals with the past and present interaction of humans with Lebanese amber, including trade and ancient uses. A section is also included on resin and copal in the Middle East and how to distinguish them from Lebanese amber.

The Early History of Lebanese Amber

It is impossible to say just who of the many peoples that inhabited the land now known as Lebanon was the first to note the existence of amber. In his work of 1865 entitled "Avienus und die Ora Maritima," W. Christ indicated that the Phoenicians were involved in active sea trading of amber from the time of Moses to about 900 B.C. There is little doubt that the Phoenicians, who occupied Lebanon as early as 3000 B.C., traded amber with neighboring lands, including Europe, and it is possible that trade began with Lebanese amber. As trade routes with northern peoples became established and the demand for amber increased, the more accessible Baltic material with larger, more workable pieces would naturally became the trade item of choice.

The question is: how much of the amber being traded originated from the Baltic area, including the Jutland peninsula, and how much was gathered from the mountains of Lebanon? While some of the Lebanese amber is solid and large enough for jewelry (Figure 4), most is brittle and occurs in small pieces. Otto Fraas, in his 1876 work "Drei Monate am Lebanon," mentions that some pieces of amber occurring in lignite in the province of Metn were "fist-sized and glass hard" and suitable for working into jewelry. He suggested that this type of material could have been traded by the Phoenicians.

The Phoenicians spoke of a gold from the mountains and called it "the shining golden substance from Lebanon" (Williamson, 1932). Since no true gold has ever been reported from there, were the Phoenicians referring to amber? Williamson (1932) reported that these traders knew about amber along the "Syrian coast." In fact, many of the reports of amber finds in western Syria probably referred to localities within the confines of present-day Lebanon, since the latter had been a republic within the borders of Syria for a number of years.

Pliny mentions amber from Syria and other Romans also noted amber from Palestine and Syria (Strong, 1966). Strong felt that "Syrian" amber may have been used locally. In the description of Roman Syria (Ta Tsin) by Kan Yang (A.D. 97) in the annals of the Han Dynasty, amber is mentioned as one of the products from Syria. At Ras Shamra (Ugarit) in Syria, amber beads were found with Mycenaean objects in the fourteenth and thirteenth centuries B. C. These beads are long, bi-conical and similar to those found in Mycenaean Greece.

In the Revelation of St. John in the Bible, there is mention of amber from Lebanon and the passage from Ezekiel 1:4, "… And a brightness was about it and out of the midst thereof as the color of amber" is thought by some to refer to Near Eastern amber. The Talmud mentions amber from Jordan (Strong, 1966). Another reference to the Hebrew use of Jordanese amber is mentioned by Rabbi Maimonides (1135-1204) when he states that "salt of Sodom, with amber of Jordan" was used as one of the components for incense burned in the services at the Tabernacle.

One of the earliest "modern" reports of Lebanese amber was its occurrence in coal beds near Beirut, a discovery noted by J. von Russegger in his 1843 book "Reisen in Europa, Asien und Afrika." The first scientific studies on Lebanese amber were conducted by John in 1876 and dealt mostly with external characteristics and investigated whether the material contained succinic acid, which at that time (and even now in some circles) was the criterion used to establish whether a source of fossilized resin was officially amber (or succinite) or was to be regarded as a retinite (lacking succinic acid). Today amber is most commonly simply defined as a hardened fossilized resin some millions of years in age. Although John (1876) reported the presence of succinic acid in both yellow transparent and yellow-brown opaque varieties of Lebanese amber, a third, red variety was found to contain both succinic and formic acid and was considered identical with a fossilized resin named Schraufiet from Austria.

Later authors challenged these findings and the results of this debate were reported by Fraas in 1878. Also in 1878, Bronner wrote a paper with additional characteristics of Lebanese amber. He determined that the red variety had a specific gravity of 1.118 while the yellow and orange varieties gave a range from 1.005 to 1.008 and confirmed that both succinic and formic acid were present in the red variety.

It is interesting that some of these early investigators discovered that Lebanese and Baltic amber had similar properties. If kauri pines were the source tree of Baltic amber, as is now suggested by chemical analysis (Poinar, 1992), then Baltic and Lebanese amber would have had similar plant origins and also similar chemical reactions even though some ninety million years separate them.

Collecting Lebanese amber

Lebanon is probably one of the most dangerous places in which to gather amber, since most of the collecting areas include localities that are highly hazardous as a result of political unrest (Poinar and Poinar, 1994). Struggles along the eastern Mediterranean have been going on for centuries, even before the ancient Phoenicians were subjugated by the Assyrian King Ashur-Nasir-Pal ll in 879 B. C. (Hitti, 1962).

So far, Raif Milki has been fortunate; none of his collecting trips have resulted in injury. The closest call he had was one day in 1985 when he was collecting amber near a militarized zone. As he was picking up small pieces of amber that had been eroded from the surrounding bedrock, he noticed a Lebanese officer with two soldiers walking through the area. He waited for them to pass and then, after collecting a few hours longer, returned to Beirut. Later that evening, when he was watching news on the television, he was shocked to hear that the one of the soldiers had stepped on a land mine and all three had perished.

After collecting amber by day, Raif spends evenings washing the material and polishing the surfaces so that a "window" can be made to expose any entrapped fossils. If an inclusion is found, the amber is carefully polished in order to better view the specimen for identification and further study. Some of the amber he has found is so fragile that it has to be immersed in liquid plastic for safe keeping (Poinar and Poinar, 1994). Each piece is then individually wrapped in soft tissue and aluminum foil, placed in an air-tight container and given an accession number for future reference.

Other resins, copals, and gums from Lebanon and the Near East

While Lower Cretaceous amber from Lebanon and the neighboring countries represents the only amber deposit in the Near East, there are many resins and copals from Mediterranean and Near Eastern plants that can be confused with amber. Historic records of the use of these resins and copals (and some gums) are well documented in many works (Moldenke and Moldenke, 1952) (Lucas, 1962). A summary of these materials with the scientific names of their source plants is presented in Table 8. While not all of these products may have been harvested in Lebanon, all at one time or another passed through Lebanon in trade, dating back to BC times. Phoenician trade with Egypt was especially profitable since the latter country used various resins for mummification and the manufacture of varnishes (Hitti, 1962; Lucas, 1962).

A word should be said concerning the definition and nature of plant resins, copals, and gums. Natural resins are water-insoluble plant products composed of complex mixtures of terpenoids, acids, and alcohols secreted from plant parenchyma cells. Gums, also natural plant secretions, differ chemically from resins and slowly dissolve in water. As resins age, they undergo oxidation and polymerization and slowly harden. When they can no longer be molded by hand, they enter a category known as copal. Most of the "resins" discussed here can technically be called "copal." Further polymerization and hardening of copal through extensive aging results in a product with the chemical and physical characteristics of amber (see Poinar, 1992 for further discussion of resins, copals, and ambers).

One of the best-known resin products is frankincense, often called the most important incense resin in the world, even today. Frankincense trees can become quite large with grand, star-shaped, greenish flowers outlined against the compound leaves. The resin is extruded as white, yellow, and tan-colored hardened drops and can be obtained by making incisions on the trunk and branches of the living trees.This material was mentioned many times in the Bible for its use in religious as well as sacrificial services of the Tabernacle and Temple (Moldenke and Moldenke, 1952). Aside from its main function as incense, frankincense was also used for embalming and even fumigating.

Herodotus (484-425 B.C.) (in Rawlinson, 1942), commented on the methods by which various resins were obtained in Arabia. In passing, he presented an intriguing account of strange creatures that "guarded" frankincense trees, thus illustrating the dangers encountered when obtaining this precious product. These creatures were described as "winged serpents, small in size, and of various colors." Herodotus mentioned that during the mating ritual of these creatures, the female seized the male and bit off his neck, but that the male was avenged for this act since at birth the young penetrated directly through the body of the female. Most scholars simply dismiss this as nonsense, but it is likely that Herodotus was recounting stories about praying mantids. These insects have sinister-appearing heads that could be construed as resembling those of serpents. They are small and range in coloration from green to brown. Their bodies are elongate, the adults are winged, and the female is notorious for biting off the male's head at the time of mating (and often consuming his entire body). The female deposits eggs in a frothy mass extruded from the posterior end of her body, which could be construed as the young emerging through her body wall. Praying mantids could have preyed on other insects in the frankincense trees, especially bees during the flowering period.

Another plant related to the frankincense is myrrh, two species of which produced the majority of the resin mentioned in ancient works, including the Bible (Moldenke and Moldenke, 1952). Myrrh is native to Arabia, Abyssinia, and the Somali coast of eastern Africa. These low thorny shrubs or small trees grow in rocky places and extrude white to tan-colored resin from their stems and branches. As with many other resins harvested by humans, incisions are made on the trunks to increase the production. Myrrh was used as incense and for embalming by the ancient Egyptians and Hebrews and this use continued during the classic Greek and Roman periods. Just when these products reached

Table 8. Plant resins, copals, and gums of the past and present from the Near East that could be confused with Lebanese amber. (Reference sources: Moldenke and Moldenke, 1952; Lucas,1962).

Product	Scientific name	Family	Common name
Balsam (balm)	Commiphora opobalsamum	Burseraceae	Balm-of-Gilead
Fir resin	Abies cilicina	Pinaceae	Cilician fir
Frankincense	Boswellia carterii	Burseraceae	Frankincense tree
Gum-arabic (gum acacia)	Acacia arabica and A. seyal	Leguminoseae	Acacia, shittah tree
Gum-tragacanth	Astralagus spp.	Leguminoseae	Milk vetch
Labdanum (ladanum, onycha)	Cistus ladaniferus and Cistus spp.	Cistaceae	Rock rose
Manna	Alhagi maurorum	Leguminoseae	Prickly alhagi
Manna	Tamarix mannifera	Tamaricaceae	Manna tamarisk
Manna	Fraxinus ornus	Oleraceae	Flowering ash
Mastic	Pistacia terebinthus (P. lentiscus)	Anacardiaceae	Mastic tree
Myrrh	Commiphora myrrha (C. kataf)	Burseraceae	Myrrh tree
Pine resin	Pinus halepensis	Pinaceae	Aleppo pine
Pine resin	Pinus pinea	Pinaceae	Stone or umbrella pine
Sandarac	Tetraclinus articulata	Cupressaceae	Sweet wood
Storax (stacte)	Styrax officinalis	Styracaceae	Storax-tree

Lebanon through trade has not been established, but the Phoenicians traded myrrh extensively (Lucas, 1962).

A resin still harvested today in Lebanon is labdanum, a product of the rock rose. This oleo-resin is soft and dark brown to black in color and emerges from the stems and leaves as a viscous exudation. Herodotus (in Rawlinson, 1942) was the first to describe the use of goats for collecting this material; the resin was gathered from their beards after they had browsed on the bushes. An alternate collecting method involved drawing leather thongs or other woven cloths over the plants during the heat of the day, when the resin was most viscous. Labdanum occurs in several species of *Cistus* including the Crete rose (*C. creticus* L.), a shrubby tree occurring along the coast as well as in the mountains of Lebanon (Nehmeh, 1978). Labdanam has a chemical structure that resembles ambergris (a product from sperm whales) and is often substituted for it in the manufacture of perfumes.

Sandarac was another resin used in Lebanon in ancient times. Although the sandarac tree is native to northern Africa, it was cultivated in Lebanon for its sweet-smelling wood, which was highly prized for cabinet making, often referred to as being worth its weight in gold. The brittle, aromatic pale yellow resin was carried by the Phoenician traders from Carthage to Lebanon, and from there to Babylon. The resin was used for making varnish and burned for incense by the Greeks and Romans (Lucas, 1962).

Conifer resin was used for various purposes, but especially for the preservation of wines and corpses. The Egyptians embalmed not only their own dead, but also a range of animals that they considered sacred (Lucas, 1962), and thus used a considerable amount of resin for mummification. Such practices resulted in an extensive trade of plant resins with neighboring countries, especially the Phoenicians (Hitti, 1962). Resin from pine and fir trees was most frequently used for embalming. The Aleppo pine, found on the dry hillsides of the lower mountain ranges, is a small tree, normally reaching only 30 feet in height; the stone pine, occurring in the sands and dunes of the littoral, rocky hills, sometimes reaches up to 70 feet in height. These two-needled pines supplied the bulk of the coniferous resin used in trade. The stone pine also had an edible seed (pinyon) which was probably a trade item. The Cilician fir, still found in Lebanon growing to a height of 90 feet and with the longest known cones of any fir, produces small, rounded lumps of ash resin (Lucas, 1962).

The plant that produces mastic or the "balm" of Genesis is a bushy tree reaching up to 10 feet in height, with pinnate evergreen leaves. The resin is a fragrant exudation from the cut stems and branches and has been an article of trade for centuries. The higher grades of yellowish-white translucent mastic are used as an astringent in medicine and in preparing jams, ice creams, and liquors. The poorer grades are used mainly for varnishes and chewing gum. The mastic tree is indigenous to the coast and lower mountains of Lebanon. Related resin-bearing trees that occur in Lebanon today are the Palestine pistachio (the turpentine tree of the Bible) and the pistachio nut tree (*Pistacia vera*).

Another product with interesting origins is manna, a term for resin from several plants as well as other materials. Manna as a tree exudate can be derived from three separate species: the prickly alhagi, the manna tamarisk, and the flowering ash. The prickly alhagi is a low, scrubby, many-branched shrub growing barely 3 feet in height. During the day, a sweet gummy exudate produced by the leaves and stems quickly hardens in the sun and is collected by shaking the bushes over a cloth. The manna tamarisk is a branching shrub or tree reaching about 15 feet in height. It produces tiny pink flowers borne in tightly packed short clusters or racemes and produces resin from wounds made by a scale insect (*Coccus manniparus*). The honey-like liquid soon dries and drops from the trees. This manna is highly regarded by the Bedouin Arabs who eat it directly or form it into cakes. This is not true honey dew since the secretion does not pass through the intestine of the scale insect (Moldenke and Moldenke, 1952). The manna ash is a tree reaching 50 feet in height with its native range from Mediterranean Europe to Lebanon and Turkey. A sweetish exudate from the limbs of this tree occurs in flake, fragment, and mass forms. It is currently used as a laxative, demulcent, and expectorant.

The resinous products of these three trees are all called manna in the Bible. Another type of manna is the gelatinous growths of *Nostoc* algae, which form rapidly during the night under heavy dews and then dry up during the day. The manna that dropped from the sky in the Old Testament has been attributed to several species of *Lecanora* lichens which when dried can be carried about by the winds and dropped to the ground in masses. These still are made into a type of bread for human consumption (Moldenke and Moldenke, 1952).

Balsam or balm is an exudation from the stems, branches, and fruits of the balm-of-Gilead tree. This small evergreen tree, which rarely reaches 15 feet in height, appears to have disappeared from Lebanon today but was cultivated at the time of Solomon. This product was used both as a spice and as incense in religious ceremonies.

Sweet storax or stacte is a product of the styrax plant, a shrub or small tree reaching 20 feet in height. The showy white flowers are quite fragrant and stand out against the dry background of the rocky Lebanese hills. The resin is extruded from incisions made in the stems and branches and was highly regarded as a perfume, although it is no longer used in commerce. A related species, *Styrax benzoin*, from southwest Asia supplies a resin known as benzoin or styrax. Another resin, known as liquid storax, is produced from the unrelated sweetgum tree (*Liquidambar orientalis*) in southwestern Asia Minor.

A well-known product from the same part of the world is gum Arabic from *Acacia* or shittah trees. Due to their great survival ability, they are one of the few remaining trees in the Arabian deserts today. The wood is burnt for fuel and charcoal, the foliage and flowers are fed to livestock, the bark is used for tanning, the fine-grained wood is used for cabinet making, and the brownish gum acacia or gum Arabic is used in the production of glues, cosmetics, pharmaceuticals, and foods.

The gum-tragacanth plant grew in the dry mountainous areas of Palestine (Moldenke and Moldenke, 1952). This dwarf shrubby plant reached only 1 to 2 feet in height and produced yellow or white pea-shaped flowers. The gum was naturally exuded from the stems and branches and is one of the oldest plant resins used by humans, dating from pre-Christian times. It is employed today as a coating and binding agent in pill manufacture.

The above plant resins (or copals) and gums can easily be confused with Lebanese amber by appearance, but can be distinguished from it by two simple steps. When placed in water, gums will start to dissolve or swell up and become spongy while amber will not be modified. When resins or copals are placed in acetone, ether, or 90% alcohol, they will immediately begin to dissolve or soften since the chemical bonds are not yet strongly cross-linked in such young material. Amber will not be modified within the first five to ten minutes but small pieces may later be affected, though to a lesser degree than the resins and copals. Any of the above-mentioned resins and gums could have insects and plant material embedded in their matrices or stuck on the outside (either naturally or intentionally). These simple tests can distinguish Lebanese amber from Near Eastern resins, copals and gums. Of course, plastic can also be used to imitate amber and distinguishing that material from amber calls for other tests.

Acknowledgments

The authors would like to thank the following persons for their assistance with the identification of the fossils discussed in this work; J. Baxter (Diptera); A. Borkent (Ceratopogonidae); D. Burckhardt (Psyllidae); J. Doyen (Coleoptera); J. Lattin (Hemiptera); J. Krantz (Acari); W. P. McCafferty (Ephemeroptera); S. O'Keefe (Coleoptera); M. Olmi (Dryinidae); G. Parsons (Coleoptera); S. Podenas (Tipulidae); M. Prentice (Hymenoptera); H. Sturm (Archeognatha); A. Wohltmann (Acari); J. Wunderlich (Araneae) and R. Zuparko (Hymenoptera). Grateful appreciation is extended to A. Haddadin for providing the fossils of *Agathis levantensis* and to John Doyen, Art Boucot, and Roberta Poinar for comments on the manuscript.

Raif Milki would like to acknowledge support given him in his amber studies by the American University of Beirut, the Al-Koura Developmental Council, Dr. Riad Tabbara, Dr. May Jurdi, and members of his family, Suzi, Nesrine, and Rania.

References

Adler, S., and O. Theodor. 1929A. The distribution of sandflies and leishmaniasis in Palestine, Syria and Mesopotamia. *Ann. Trop. Med. Parasitology* 23: 269-306.

Adler, S., and O. Theodor. 1929B. Observations of *Leishmania ceramondactyli* n. sp. *Trans. R. Soc. Trop. Med. Hyg.* 22: 343-56.

Auezova G., Z. Brushko, and R. Kubykin. 1990. Feeding of biting midges (Leptoconopidae) on Reptiles. Abst. 2nd Int. Congr. Dipterology, Bratislava, Czechoslovakia . 324 pp.

Azar, D., A. Nel, M. Solignac, J.-C. Paicheler, and F. Bouchet. 1999A. New genera and species of psychodoid flies from the Lower Cretaceous amber of Lebanon. *Palaeontology* 42: 1101-36.

Azar, D. , G. Fleck, A. Nel, and M. Solignac. 1999B. A new enicocephalid bug, *Enicocephalinus acragrimaldi* gen. nov., sp. nov.from the Lower Cretaceous amber of Lebanon (Insects, Heteroptera, Enicocephalidae). Est. Mus. Cienc. Nat. de Alava (1999) 14 (Num. Espec. 2): 217-30.

Azar, D. A. Nel, and M. Solignac. 2000. A new Coniopterygidae from Lebanese amber. *Acta Geologica Hispanica* 35: 31-36.

Bandel, K., and A. Haddadin. 1979. The depositional environment of amber-bearing rocks in Jordan. *Dirasat* 6: 39-62.

Bandel, K., and N.Vavra. 1981. Ein fossiles Harz aus der Unterkreide Jordaniens. *N. Jb. Geol. Palaeont Mh.*, 1981 (1): 19-33.

Benton, M.J. 1990. Red Queen Hypothesis. In: *Paleobiology, A Synthesis*. Eds. D. E. G. Briggs and P. Crowther. Blackwell Science Ltd., Oxford. pp. 119-24.

Borkent, A. 1995. *Biting Midges in the Cretaceous Amber of North America (Diptera: Ceratopogonidae)*. Backhuys Pub., Leiden. 237 pp.

Borkent, A. 2000. Biting midges (Ceratopogonidae: Diptera) from Lower Cretaceous Lebanese amber with a discussion of the diversity and patterns found in other ambers. In: *Studies on Fossils in Amber with Particular Reference to the Cretaceous of New Jersey*. Ed. D. Grimaldi. Backhuys Publishers, Leiden. pp. 355-451.

Borkent, A. 2001. *Leptoconops* (Diptera: Ceratopogonidae), the earliest extant lineage of biting midge, discovered in 120-122 million-year-old Lebanese amber. *Amer. Mus. Novitates* 3328: 1-11.

Borkent, A., W. W. Wirth, and A. L. Dyce. 1987. The newly discovered male of *Austroconops* (Ceratopogonidae: Diptera) with a discussion of the phylogeny of the basal lineages of the Ceratopogonidae. *Proc. Entomol. Soc. Wash.* 89: 587-606.

Borror, D. J., C. A. Triplehorn, and N. F. Johnson. 1989. *Introduction to the Study of Insects*. 6th edition. Saunders College Publishing, Philadelphia. 875 pp.

Boucot, A. J. 1990. *Evolutionary Paleobiology of Behavior and Coevolution*. Elsevier, Amsterdam. 725 pp.

Brandão, C. R. F., R. G. Martins-Neto, and M. A. Vulcano. 1989. The earliest known fossil ant (first southern hemisphere Mesozoic record)(Hymenoptera: Formicidae: Myrmeciinae). *Psyche* 96: 195-208.

Bronner, H. 1878. Über einige fossile Harze vom Libanon. Jahreshefte des Vereins für vaterländische Naturkunde in Württemberg 34: 81-90.

Brown, J. H., and M. V. Lomolino. 1998. *Biogeography* (second edition). Sinauer Associates, Inc. Pub., Sunderland, MA. 691 pp.

Brundin, L. 1976. A Neocomian chironomid and Podonominae-Aphroteniinae (Diptera) in the light of phylogenetics and biogeography. *Zool. Scripta* 5: 139-60.

Cano, R. J., H. N. Poinar, N.J. Pieniazek, A. Acra, and G. O. Poinar, Jr. 1993. Amplification and sequencing of DNA from a 120-135 million year old weevil. *Nature* 363: 536-38.

Cleveland, L. R. 1934. The wood-feeding roach *Cryptocercus*, its protozoa, and the symbiosis between protozoa and roach. *Mem. Amer. Acad. Arts Sciences* 17: 342 pp. (+ 60 plates).

Cookson, I. C., and S.L. Duigan. 1951. Tertiary Araucariaceae from South-Eastern Australia, with notes on living species. Australian J. Sci. Res. 4:415-49.

Crowson, R. A. 1981. *The Biology of the Coleoptera*. Academic Press, New York. 802 pp.

Daly, H. V., J. T. Doyen, and A. H. Purcell. 1998. *Introduction to Insect Biology and Diversity* (second edition). Oxford Univeristy Press, Oxford. 680 pp.

Dolling, W. R. 1981. A rationalized classification of the burrower bugs (Cydnidae). *Systematic Entomol.* 6: 61-76.

Downes, J. A. 1971. The ecology of blood-sucking Diptera: An evolutionary prespective. In: *Ecology and Physiology of Parasites: A Symposium*. Ed. A. M. Fallis. University of Toronto Press, Toronto. pp. 232-58.

Edwards, W. N. 1929. Lower Cretaceous plants from Syria and Tansjordania. *Ann. Mag. Natural History* 4: 394-405.

Eldredge, N. 1985. *Time Frames*. Simon and Schuster, Inc., New York. 240 pp.

Fennah, R. G. 1987. A new genus and species of Cixiidae (Homoptera: Fulgoroidea) from Lower Cretaceous amber. *J. Natural History* 21: 1237-40.

Fraas, O. 1876. Drei Monate am Libanon. Levig and Miller, Stuttgart.

Fraas, O. 1878. Geologisches aus dem Lebanon. Jahreshefte des Vereins für vaterländische Naturkunde in Württemberg 34: 257-391.

Frickhinger, K. A. 1991. *Fossil Atlas Fishes*. English translation. Hans A. Baensch Pub., Melle, Germany.

Gough, L. J., and J. S. Mills. 1972. The composition of succinite (Baltic amber). *Nature* 239: 527-28.

Goulet, H., and J. T. Huber. 1993. *Hymenoptera of the World: An Identification Guide to Families*. Publication 1894/E, Research Branch, Agriculture Canada, Ottawa. 668 pp.

Graffham, A. A. 2000. Catalog of fossils. Geological Enterprises Bull. 52: 1-63.

Grimaldi, D. 2000. A diverse fauna of Neuropterodea in amber from the Cretaceous of New Jersey. In: *Studies on Fossils in Amber with Particular Reference to the Cretaceous of New Jersey*. Ed. D. Grimaldi. Backhuys Pub., Leiden. 259-303.

Grimaldi, D., and J. Cumming. 1999. Brachyceran Diptera in Cretaceous ambers and Mesozoic diversification of the Eremoneura. *Bull. Amer. Mus. Nat. Hist.* 239: 1-124.

Halliday, T. R., and K. Adler (Eds.). 1986. *The Encyclopedia of Reptiles and Amphibians*. Facts on File, New York. 143 pp.

Hanson, P. E., and I. D. Gauld. 1995. *The Hymenoptera of Costa Rica*. Oxford University Press, Oxford.

Harland, W. B., R. L. Armstrong, A. V. Cox, L. E. Craig, A. G. Smith, and D. G. Smith. 1990. *A Geologic Time Scale*. Cambridge University Press., Cambridge. 263 pp.

Heie, O., and D. Azar. 2000. Two new species of aphids found in Lebanese amber and a revision of the family Tajmyaphididae Kononova, 1975 (Hemiptera: Sternorrhyncha). *Ann. Entomol. Soc. Amer.* 93: 1222-25.

Hennig, W. 1970. Insektenfossilien aus der untered Kreide. II. Empedidae (Diptera: Brachycera). *Stuttgarter Beitr. Naturk.* 214: 1-12.

Hennig,W. 1971. Insektenfossilien aus der unteren Kreide. III. Empidiformia ("Microphorinae") aus der unteren Kreide und aus dem Baltischen Bernstein: ein Vertreter des Cyclorrhapha aus der unteren Kreide. *Stuttgarter Beitr. Naturk.* (Serie A) 232: 1-28.

Hennig, W. 1972. Insektenfossilien aus der unteren Kreide. IV. Psychodidae (Phlebotominae), mit einer kritischen Ubersicht uber das phylogenetische System der Familie und die bisher beschriebenen Fossilien. *Stuttgarter Beitr. Naturk.* (Serie A) 241: 1-69.

Hilts, P. J. 1996. Expedition to far New Jersey finds trove of amber fossils. *The New York Times,* January 30, 1996, p. B5.

Hitti, P.K. 1962. *Lebanon in History.* MacMillan and Co., Ltd. London. 548 pp.

Howden, H. F., and R. I. Storey. 1992. Phylogeny of the Rhyparini and the new tribe Stereomerini, with descriptions of new genera and species (Coleoptera; Scarabaeidae; Aphodiinae). *Can. J. Zool.* 70: 1810-23.

Jacobs, L. 1993. *Quest for the African Dinosaurs.* Villard Books, New York. 314 pp.

Jell, P. A., and P. M. Duncan. 1986. Invertebrates, mainly insects, from the freshwater, lower Cretaceous, Koonwarra Fossil Bed (Korumburra Group), South Gippsland, Victoria. *Mem. Ass. Austral. Paleontols.* 3: 111-205.

John, K. 1876. Bernstein und Schraufit aus dem Libanon. *Verhandlungen k.k. geologische Reichsanstalt* (Wien) 1876: 255-57.

Kettle, D. S. 1984. *Medical and Veterinary Entomology.* John Wiley and Sons, New York. 658 pp.

Koteja, J. 2000. Advances in the study of fossil coccids (Hemiptera: Coccinea). *Polish J. Entomol.* 69: 187-218.

Kuschel, G., and G. O. Poinar, Jr. 1993. *Libanorhinus succinus* gen. and sp. n. (Coleoptera: Nemonychidae) from Lebanese amber. *Ent. Scand.* 24: 143-46.

Labandeira, C. C., B. S. Beall, and F. M. Hueber 1988. Early insect diversification: evidence from a Lower Devonian bristletail from Quebec. *Science* 242: 913-16.

Labandeira, C. C., and J. J. Sepkoski, Jr. 1993. Insect diversity in the fossil record. *Science* 261: 310-15.

Lainson, R., and J. J. Shaw. 1987. Evolution, classification and geographical distribution. In: *The Leishmaniases in Biology and Medicine.* Vol.1. Eds. W. Peters and R. Killick-Kendrick. Academic Press, London. pp.13-112.

Lambert, J. B., J. S. Frye, and G. O. Poinar, Jr. 1990. Analysis of North American amber by carbon-13 NMR Spectroscopy. *Geoarchaeology* 5: 43-52.

Lambert, J. B., S. C. Johnson, and G. O. Poinar, Jr. 1996. Nuclear magnetic resonance characterization of Cretaceous amber. *Archaeometry* 38: 325-35.

Langenheim, J. H. 1969. Amber: a botanical inquiry. *Science* 163: 1157-69.

Langenheim, J. H., and C. W. Beck. 1968. Catalogue of infrared spectra of fossil resin (ambers) 1. North and South America. *Botanical Museum Leaflets,* Harvard University 22: 65-120.

Larsen, T. B. 1974. *Butterflies of Lebanon.* National Council for Scientific Research, Beirut. 255 pp.

de Laubenfels, D. J. 1988. *Agathis.* Flora Malesiana 103: 429-42.

Lillegraven, J. A., Z. Kielan-Jaworowska, and W. A. Clemens (Eds.). 1979. *Mesozoic Mammals.* University of California Press, Berkeley. 311 pp.

Lucas, A. 1962. *Ancient Egyptian Materials and Industries* (fourth edition revised and enlarged by T. R. Harris). Edward Arnold Ltd., London.

McAlpine, J. F. (Ed.) 1981. Manual of Nearctic Diptera. Vol. 1, Monograph No.27, Research Brance, Agriculture Canada, Ottawa, 1-674.

McAlpine, J. F. (Ed.) 1987. Manual of Nearctic Diptera, Vol ll, Monograph No. 28, Research Branch, Agriculture Canada, Ottawa, 675-1332.

McCafferty, W. P. 1997. Discovery and analysis of the oldest mayflies (Insecta, Ephemeroptera) known from amber. *Bull. Soc. Hist. Nat. Toulouse* 133: 77-82.

McKeever, S. 1977. Observations of *Corethrella* feeding on tree frogs (*Hyla*). *Mosquito News* 37: 522-23.

Meinander, M. 1975. Fossil Coniopterygidae (Neuroptera). *Notulae Ent.* 55: 53-57.

Moldenke, H. N., and A. L. Moldenke. 1952. *Plants of the Bible.* The Ronald Press Co., N.Y. 328 pp.

Nehmeh, M. 1978. *Wild Flowers of Lebanon.* National Council for Scientific Research, Beirut. 238 pp.

Newman, W. A. 1996. Origins of Southern Hemisphere endemism, especially among marine Crustacea. *Mem. Queensland Museum* 31: 51-76.

Nissenbaum, A., and A. Horowitz. 1992. The Levantine amber belt. *J. African Earth Sciences* 14: 295-300.

Olmi, M. 1997. New Oriental and Neotropical Dryinidae (Hymenoptera Chrysidoidea). *Frustula Entomologica* 32: 152-67.

Olmi, M. 1998. New fossil Dryinidae from Baltic and Lebanese amber (Hymenoptera Chrysidoidea). *Frustula Entomologica* 34: 48-67.

Paul, C. R. C. 1988. Extinction and survival in the echinoderms. In: Extinction and Survival in the Fossil Record. Ed. G. P. Larwood. Clarendon Press, Oxford. pp. 155-70.

Pike, E. M. 1994. Historical changes in insect community structure as indicated by Hexapods of upper Cretaceous Alberta (Grassy Lake) amber. *Canadian Entomol.* 126: 695-702.

Pike, E. M. 1995. Amber Taphonomy and the Grassy Lake, Alberta, amber fauna. PhD. Dissertation, The University of Calgary, Alberta.

Podenas, S., G. Poinar, Jr., and R. Milki. 2001. New crane flies (Diptera: Limoniidae) from Lebanese amber. *Proc. Entomol. Soc. Wash.* 103: 433-36.

Poinar, G. O., Jr. 1990. Insects and plant pathogens attacking kauri pines (*Agathis* spp.). Unpublished report. 12 pp.

Poinar, G. O., Jr. 1992. *Life in Amber.* Stanford University Press, Stanford, CA. 350 pp.

Poinar, G. O., Jr. 2001. Nematoda and Nematomorpha. In: *Ecology and Classification of North American Freshwater Invertebrates* (second edition). Eds. J. H.Thorp and A. P. Covich. Academic Press, N.Y. pp. 255-95.

Poinar, G. O., Jr., and J. Havercamp. 1985. Use of pyrolysis mass spectrometry in the identification of amber samples. *J. Baltic Studies* 16: 210-21.

Poinar, G. O., Jr., and R. Poinar 1994. *The Quest for Life in Amber.* Addison Wesley, Reading, MA. 219 pp.

Poinar, G. O., Jr., A. Acra, and F. Acra. 1994A. Earliest fossil nematode (Mermithidae) in Cretaceous Lebanese amber. *Fundam. Appl. Nematol.* 17: 475-77.

Poinar, G. O., Jr., A. Acra, and F. Acra. 1994B. Animal-animal parasitism in Lebanese amber. *Med. Sci. Res.* 22: 159.

Poinar, G. O., Jr., and L. A. Stange. 1996. A new antlion from Dominican amber (Neuroptera: Myrmeliontidae). *Experientia* 52: 383-86.

Poinar, G. O., Jr., and R. Poinar 1999. *The Amber Forest.* Princeton University Press. 238 pp.

Poinar, G. O., Jr., B. Archibald, and A. Brown. 1999. New amber deposit provides evidence of early Paleogene extinctions, Paleoclimates and past distributions. *Canadian Entomol.* 131: 171-77.

Poinar, G. O., Jr., C. Baroni Urbani, and A. Brown. 2000. The oldest ants are Cretaceous, not Eocene. Reply. *Canadian Entomol.* 132: 695-96.

Poinar, G. O., Jr., T. J. Zavortink, T. Pike, and P. A. Johnson. 2000. *Paleoculicis minutus* (Diptera: Culicidae) n. gen., n.sp., from Cretaceous Canadian amber, with a summary of described fossil mosquitoes. *Acta Geol. Hispan.* 35: 119-28.

Ponomarenko, N. G. 1981. New Hymenoptera of the family Dryinidae of the Upper Cretaceous of Taimyr and Canada. *Paleon. Zhurnal* 1981: 139-43.

Prentice, M. 1993. Early Cretaceous Aculeata from Lebanese amber. *Sphecos* 26: 8.

Prentice, M. 1994. Some further notes on Lebanese Aculeata. *Sphecos* 27: 12.

Prentice, M. A. ,G. O. Poinar, Jr., and R. Milki. 1996. Fossil scolebythids (Hymenoptera: Scolebythidae) from Lebanese and Dominican amber. *Proc. Entomol. Soc. Wash.* 98: 802-11.

Rawlinson, G. 1942. *Herodotus: The Persian Wars* (English translation). The Modern Library, NY. 714 pp.

Roth, B., G. O. Poinar, Jr., A.Acra, and F. Acra 1996. Probable pupillid land snail of early Cretaceous (Hauterivian) age in amber from Lebanon. *Veliger* 39: 78-79.

Russegger, J. von. 1843. Reisen in Europa, Asien und Afrika. Vol.1, E. Schweizerbart'sche, Stuttgart.

Savage, R. J. G. 1988. Extinction and the fossil mammal record. In: *Extinction and Survival in the Fossil Record.* Ed. G. P. Larwood. Clarendon Press, Oxford. pp. 319-34.

Schlee, D. 1970. Verwandtschaftsforschung an fossilen und rezenten Aleyrodina (Insecta, Hemiptera). *Stuttgarter Beitr. naturk.* (Serie B) 213: 72 pp.

Schlee, D. 1972. Bernstein aus dem Libanon. *Kosmos* (Stuttgart) 68: 460-63.

Schlee, D. 1973. Harzkonservierte Vogelfedern aus der untersten Kreide. *J. Ornithol.* 114: 207-19.

Schlee, D. 1990. Das Bernstein-Kabinett. *Stuttgarter Beitr. Naturk.* (Serie C) 28: 100 pp.

Schlee, D., and H.-G. Dietrich. 1970. Insektenfuhhender Bernstein aus der Unterkreide des Libanon. *Neues Jahrbuch Geol. Palaontol.* 1970 (1): 40-50.

Schlee, D., and W. Glöckner. 1978. Bernstein. *Stuttgarter Beitr. Naturk.* (Serie C) 8: 72 pp.

Schuh, R. T., and J. A. Slater. 1995. *True Bugs of the World (Hemiptera: Heteroptera).* Cornell University Press, Ithaca, NY. 336 pp.

Serhal, A. 1985. *Wild Mammals of Lebanon.* Rihani House Est., Beirut. 151 pp. (in Arabic)

Shinaq, R., and K. Bandel 1998. The flora of an estuarine channel margin in the Early Cretaceous of Jordan. *Freiberger Forschungsheft.* 474: 39-57.

Smith, A. G., D. G. Smith, and B. M. Funnell. 1994. *Atlas of Mesozoic and Cenozoic Coastlines.* Cambridge Univ. Press, Cambridge.

Spahr, U. 1992. Erganzungen und Berichtigungen zu R. Keilbachs Bibloigraphie und Liste der Bernsteinfossilien -Klass Insecta. *Stuttgarter Beitr. Natur.* Ser. B. Nr. 182: 1-102.

Stenseth, N. C., and J. Maynard Smith 1984. Coevolution in ecosystems: Red Queen evolution or stasis? *Evolution* 38: 870-80.

Strong, D. E. 1966. Catalogue of the carved amber in the Department of Greek and Roman antiquities. The Trustees of the British Museum, London. 104 pp.

Sturm, H., and G. O. Poinar, Jr. 1998. *Cretaceomachilis libanensis,* the oldest known Bristle-tail of the family Meinertellidae (Machiloidea, Arachaeognatha, Insecta) from the Lebanese amber. *Mitt. Mus. Nat. kd. Berl. Dtsch. entomol. Zeit.* 45: 43-48.

Szadziewski, R. 1995. The oldest fossil Corethrellidae (Diptera) from Lower Cretaceous Lebanese amber. *Acta. zool.cracov.* 38: 177-81.

Szadziewski, R. 1996. Biting midges from the Lower Cretaceous amber of Lebanon and Upper Cretaceous Siberian amber of Taimyr (Diptera, Ceratopogonidae). *Studia dipterologica* 3: 23-86.

Taylor, T. N., and E. L. Taylor. 1993. *The Biology and Evolution of Fossil Plants.* Prentice Hall, Englewood Cliffs, New Jersey. 982 pp.

Thomas, B. R. 1969. Kauri resins—modern and fossil. In: *Organic Geochemistry*, G. Eglinton and M. Murphy, eds. Springer-Verlag, Berlin, pp. 599-618.

Tidwell, W. D. 1998. *Common Fossil Plants of Western North America.* Smithsonian Institution Press, Washington, D.C., 299 pp.

Ting, W. S., and A. Nissenbaum. 1986. Fungi in lower Cretaceous amber. Special publication, Exploration and Developmental Research Center, Chinese Petroleum Corporation, Miaoli, Taiwan. 27 pp.

Unwin, D. M. 1988. Extinction and survival in birds. In: *Extinction and Survival in the Fossil Record.* Ed.G. P. Larwood. Clarendon Press, Oxford. pp. 295-318.

Van Valen, L. M. 1973. A new evolutionary law. *Evolutionary Theory* 1: 1-30.

Weishampel, D. B., P. Dodson, and H. Osmolska (Eds.). 1990. *The Dinosauria.* University of California Press, Berkeley, CA. 733 pp.

Whalley, P. 1978. New taxa of fossil and recent Micropterigidae with a discussion of their evolution and a comment on the evolution of Lepidoptera (Insecta). *Ann. Transvaal Mus.* 31: 65-81.

Whalley, P. 1980. Neuroptera (Insecta) in amber from the Lower Cretaceous of Lebanon. *Bull. Brit. Mus. Nat. Hist. (Geol.)* 33: 157-64.

Whalley, P. 1981. Insects from Lebanese amber. Unpublished report, British Mus.Nat. Hist. (Geol.). 11 pp.

White, M. E. 1981. Revision of the Talbragar fish bed flora (Jurassic) of New South Wales. *Records Australian Mus.* 33: 695-721.

White, R. E. 1983. *A Field Guide to the Beetles of North America.* Houghton Mifflin Co., Boston. 368 pp.

Whitmore, T. C. 1980. Monograph of *Agathis. Plant Systematics and Evolution* 135: 41-69.

Wichard, W., and W. Weitschat. 1996. Wasserinsekten im Bernstein. *Entom. Mitt.* 4: 1-121.

Williams, J. A., and J. D. Erdman. 1968. Occurrence of blood meals in two species of *Corethrella* in Florida. *Ann. Entomol. Soc. Am.* 61: 1336.

Williamson, G. C. 1932. *The Book of Amber.* Ernest Benn Ltd., London. 268 pp.

Wirth, W. W. 1956. New species and records of biting midges ectoparasitic on insects (Diptera, Heleidae). *Ann. Entomol. Soc. America* 49: 356-64.

Wunderlich, J., and R. Milki. 2001. Description of the first spider in Cretaceous Lebanese amber (Aranaeae: Dysdoidea: Oonopidae). *Beitr. Araneol.* 3: (in press)

zur Strassen, R. 1973. Fossile Fransenflugler aus mesozoischem Bernstein des Lebanon. *Stuttgart Beitr. Naturk.* (Serie A) 267: 1-51.

APPENDIX

Description of *Agathis levantensis* sp. n.
(Figures 5-9, 15)

Order Coniferales
Family Araucariaceae
Genus Agathis Salisbury 1807

The description of this new species is based on leaf remains, female cone scales, and pollen grains recovered in sedimentary rocks from the amber-bearing beds of Jordan, as well as wood analysis previously reported and chemical analysis of the amber performed by two separate laboratories.

1) Leaves narrow to broadly lanceolate with rounded apex; shape of base not clear since all bases were either broken away or had disintegrated. The most complete leaves range from 3 –to 5 cm in length and 0.8 –to 1.0 cm in width at the widest point; margins entire, flat, and not appreciably thickened; venation consists of numerous parallel, fairly inconspicuous veins (Figures.5-7), without prominent midrib. A microscopic examination of these leaf remains shows veins (Figure 7) as well as leaf cells (Figure 8). The leaf size falls within the range of both extant and fossil species of *Agathis* (Cookson and Duigan, 1951). Leaves identified as *Agathis* were previously reported from these deposits by Bandel and Haddadin (1979). Also in his study of the Lower Cretaceous plants of Jordan, Edwards (1929) identified leaves of *Brachyphyllum* and *Podozamites* in a clay bed near the Zerka Valley some 12 km southeast of Beirut. Many earlier described species of *Podozamites* are now suspected to belong to the genus *Agathis* (Tidwell, 1998; White, 1981). It is possible that the leaf fossils assigned to these genera by Edwards (1929) belonged to *Agathis levantensis* .

2) Structures that resemble female cone scales of present-day *Agathis* were also recovered from the same beds that contained the leaves (Figure 9). These scales were 2 cm broad with lateral projections and a slightly crenulate terminus (not pointed as in *Araucaria* cone scales).

3) Pollen grains found in macerated leaf tissue of *A. levantensis* were characteristic of both extant and fossil members of *Agathis*. These grains were spherical and non-aperturate and had granular exines (Figure 15). They ranged in size from 35 –to 40 µm. Pollen grains of extant

Agathis range from 40 to 56 μm in diameter and the average diameter of pollen of the fossil *A. yallournensis* was 46 μm (Cookson and Duigan, 1951). The exine of *Agathis* pollen is composed of small, closely arranged granules, each terminating in a minute spine (Figure 15). Fossil grains are usually flattened and frequently crumpled and fragmentary (Cookson and Duigan, 1951), which was the case in the present study. While pollen grains of extant *Agathis* and *Araucaria* are similar, the average diameter of *Araucaria* pollen is greater (59-81 μm) than that of *Agathis* (40-56 μm) (Cookson and Duigan, 1951).

4). Analysis of wood, some of which was still associated with amber fragments, from the Lower Cretaceous beds in Jordan where the *Agathis* leaves were recovered, showed anatomical features characteristic of present-day araucarians (Bandel and Haddidin, 1979).

5) Chemical analysis of the amber found in close proximity to the leaves showed a similarity to fossilized resin from known populations of *Agathis* (Bandel and Haddadin, 1979; Lambert et al., 1996). The spectra of Lebanese and Jordanian amber were identical with NMR analysis, thus indicating that both were part of the Lower Cretaceous araucarian forest that extended over much of the Levant (Lambert et al., 1996). The type specimen of *A. levantensis* is the leaf shown in Figure 6 (Accession No. OA-11-92), which is deposited in the Poinar amber collection maintained at Oregon State University. This is the first fossil *Agathis* described from the northern hemisphere and can be separated from known extant and extinct species by the size of the pollen grains, the size and shape of the leaves, and characters of the female cone scales. Previous fossil *Agathis* were described from the Tertiary, Cretaceous, and Jurassic of Australia and New Zealand (Cookson and Duigan, 1951; White, 1981) .

Index